THE FRANKLIN INSTITUTE PRESS℠

APPLICATIONS OF INDUCTIVELY COUPLED PLASMAS TO EMISSION SPECTROSCOPY

R. M. Barnes, editor

1977 Eastern Analytical Symposium
M. W. Miller, editor

THE FRANKLIN INSTITUTE PRESS
Philadelphia, Pennsylvania

Current printing (last digit):

5 4 3 2
ISBN Number: 0-89168-020-9
Library of Congress Catalog Card Number: 78-70143
Printed in the United States of America.

PREFACE

The Eastern Analytical Symposium, E.A.S. is privileged to publish a series of books based on the symposia presented at the Seventeenth Eastern Analytical Symposium. This series is a continuation of an Eastern Analytical Symposium policy to publish the symposia of greatest current interest. Some papers consist of reviews of the state of the art with respect to specific techniques, while others represent the results of applying techniques to particular problems.

The series editor has grouped together symposia and papers on similar topics. The individual books are edited by the symposium session chairman or in the case of multi-session symposia, the topic coordinator. Thanks are due to the chairmen of the sessions in which the papers were presented, since they were ultimately responsible for the selection of speakers, under the overall guidance of G. W. Ewing and B. Bulkin as Program Chairmen.

The series editor wishes to extend his thanks to A. Hartkopf who helped prepare the indexes.

Michael W. Miller - Series Editor

CONTRIBUTORS

F.N. ABERCROMBIE, Montana Bureau of Mines and Geology, Montana College of Mineral Science and Technology, Butte, MT 59701

RAMON M. BARNES, Department of Chemistry, GRC Tower I, University of Massachusetts, Amherst, MA 01003

A.R. BARRINGER, Barringer Research Limited, 304 Carlingview Drive, Rexdale, Ontario, Canada M9W 5G2

KENNETH W. BOYER, Division of Chemistry and Physics, US Food and Drug Administration, 200 C Street SW, Washington, DC 20204

RICHARD F. BROWNER, School of Chemistry, Georgia Institute of Technology, Atlanta, GA 30332

A.J. ELLGREN, Union Carbide Corporation, Tarrytown Technical Center, Tarrytown, NY 10591

DANOLD GOLIGHTLY, US Geological Survey, National Center, 12201 Sunrise Valley Drive, Reston, VA 22092

J. BENTON JONES, JR., Department of Horticulture, University of Georgia, Athens, GA 30602

JOHN W. JONES, Division of Chemistry and Physics, US Food and Drug Administration, 200 C Street SW, Washington, DC 20204

R.J. LEWANDOWSKI, Union Carbide Corporation, Tarrytown Technical Center, Tarrytown, NY 10591

A.D. MURRAY, Barringer Research Limited, 304 Carlingview Drive, Rexdale, Ontario, Canada M9W 5G2

JOHN A. NORRIS, Analytical Chemistry Division, Institute for Materials Research, National Bureau of Standards, Washington, DC 20234

R.G. PARKHURST, Union Carbide Corporation, Tarrytown Technical Center, Tarrytown, NY 10591

M.D. SILVESTER, Barringer Research Limited, 304 Carlingview Drive, Rexdale, Ontario, Canada M9W 5G2

ROBERT L. WATTERS, JR., Analytical Chemistry Division, Institute for Materials Research, National Bureau of Standards, Washington, DC 20234

W. ZAMECHEK, Union Carbide Corporation, Tarrytown Technical Center, Tarrytown, NY 10591

CONTENTS

INTRODUCTION

Inductively coupled plasma atomic emission spectroscopy (ICP-AES) is becoming a well-known and widely applied technique in analytical chemistry. With the number of ICP-AES instruments in the world capable of multielement spectrochemical analysis rapidly approaching 200, the organizers of the Seventeenth Eastern Analytical Symposium suggested holding a special symposium to report some recent and novel applications of the ICP-AES technique. A number of experienced ICP users agreed to participate in the symposium, which was held at New York's Americana Hotel on November 30, 1977. This monograph contains the eight invited ICP-AES applications papers. An introductory review sets the tone, and the topics cover: the practical analysis of metals, soil extracts, plant tissue ash, geochemical samples, air particulates, and foods; the approaches to sampling employed for ICP-AES analyses; and definition of sources of inaccuracy and imprecision in ICP-AES.

The interested reader should find this collection of original papers a useful introduction to this fascinating new field of analytical chemistry.

Ramon M. Barnes
Amherst, Massachusetts

REVIEW OF THE APPLICATIONS OF THE INDUCTIVELY COUPLED PLASMA

Ramon M. Barnes

INTRODUCTION

In today's modern spectrochemical laboratory, the analyst has a choice of diverse emission, absorption, and fluorescence spectrochemical techniques from which to select in solving the particular problem of current interest. Combustion flame emission and absorption, especially combined with electrothermal atomizers, x-ray fluorescence, and electrically excited discharge sources are presently the most popular spectrochemical techniques. The selection of one of the electrically excited discharge sources from Table I can be based upon the sample form, the time required, and the performance needed for the analysis. Excitation sources operated at atmospheric pressure generally provide convenient sample manipulation for solids, liquids, or powders, although a commercial glow discharge using solid samples is almost as conveniently applied. In

TABLE I. ELECTRICAL EXCITATION SOURCES

Pressure	Electrical Conditions Steady Current	Transient Current	High Frequency	Microwave Frequency
Atmospheric	DC ARC	SPARK	ICP	CCP
	AC ARC	EXPLODING	CCP	MIP
	PLASMA JET	WIRE, FOIL		
Reduced	HOLLOW CATHODE	FLASH LAMP	EDL	EDL
	GLOW DISCHARGE		ICP LAMP	MIP
	GAS LASER		CCP	
Vacuum		SPARK		

ICP = inductively coupled plasma; CCP = capacitatively coupled plasma; MIP = microwave coupled plasma; EDL = electrodeless discharge lamps

the dozen years since the first published applications of the inductively coupled plasma discharge (ICP) for spectrochemical analysis, experimental evidence has accumulated to demonstrate that the analytical performance of the ICP can be superior to that of alternative excitation sources, such as combustion flames or microwave and dc plasmas, for many single and multielement determinations. When applied for practical emission analysis of solutions, the ICP discharge is a proven excellent tool.

The availability in recent years of commercial ICP-atomic emission spectroscopy (AES) instrument systems has spurred increasingly widespread utilization of ICP sources in routine as well as specialized applications. Capabilities of ICP-AES challenge the analyst to undertake new and difficult sample problems and to achieve analyses previously beyond reach with conventional emission excitation sources or other spectrochemical techniques. This article and the seven following ones highlight significant and productive fields of applications of ICP-AES and illustrate numerous general and unique uses to which the ICP has been put in practice. Although this survey is not comprehensive, it should serve to stimulate further reading in key publications dealing with ICP-AES. Among these are reviews (1-4), numerous publications, and the popular *ICP Information Newsletter* (5).

ICP-AES PROPERTIES AND CAPABILITIES

Instrumentation

The inductively coupled plasma discharge is generated by inducing a magnetic field in a flowing, conducting gas, usually argon and sometimes argon and nitrogen, by means of a water-cooled copper coil, which surrounds quartz or boron nitride confinement tubes through which the gas flows. A schematic representation of a typical ICP arrangement is given in Figure 1 and indicates the general alignment of three coaxial and concentric quartz tubes (6). Two and

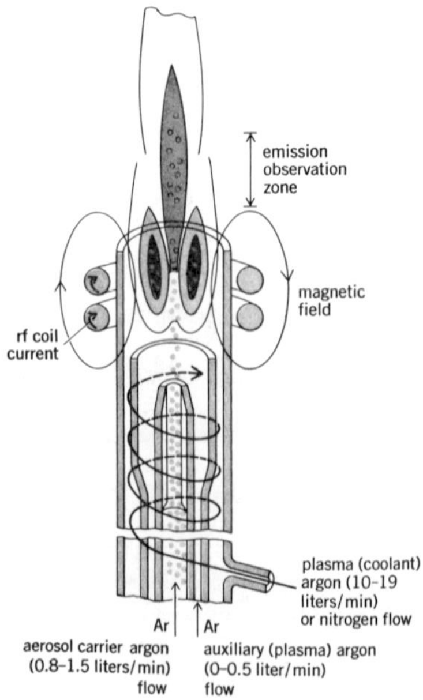

FIGURE I. SCHEMATIC DIAGRAM OF AN ICP (6)

sometimes three gas streams are used to sustain the high-temperature discharge. Important features of the ICP discharge include the formation of the plasma which reaches temperatures of the order of 10000°K in an inert argon atmosphere without contact with and contamination from electrodes. Not only do the gas flows sustain the high-temperature discharge, but they position it relative to the walls and the induction coil, prevent the outer confinement tube walls from melting or ablating, and form the discharge into an annular shape through which sample aerosol can be transported by the sample aerosol carrier gas flowing from the central tube. Although a variety of torch configurations have been used, only a few have become popular and widespread(3). One of these operates with an inner diameter of about 18 mm and a main body length ranging from 3 to 6 inches. The positioning and concentricity of these precision-ground quartz tubes is a critical fabrication parameter in high-performance ICP instruments. The outermost annulus of the torch geometry carries either argon or, in high-power discharges, nitrogen at the highest of the three gas flow rates, generally 10 to 20 L Ar/min. This plasma gas flow is responsible for the maintenance of the discharge and the insulation of the tube walls. The sample is transported as an aerosol through the 1.0-mm to 1.5-mm diameter orifice at argon flow rates from 0.5 to about 1.5 L/min, depending upon the nebulizer and corresponding to solution uptake rates of 1 to 3 mL/min. The third auxiliary gas flow is optional and used at rates generally less than 1 L Ar/min in particular applications such as in the analysis of organics.

As illustrated in Figure 2, samples in the form of liquids, solids, powders and gas have been introduced into the ICP discharge(7). However, the most common technique is to nebulize solutions pneumatically or ultrasonically using

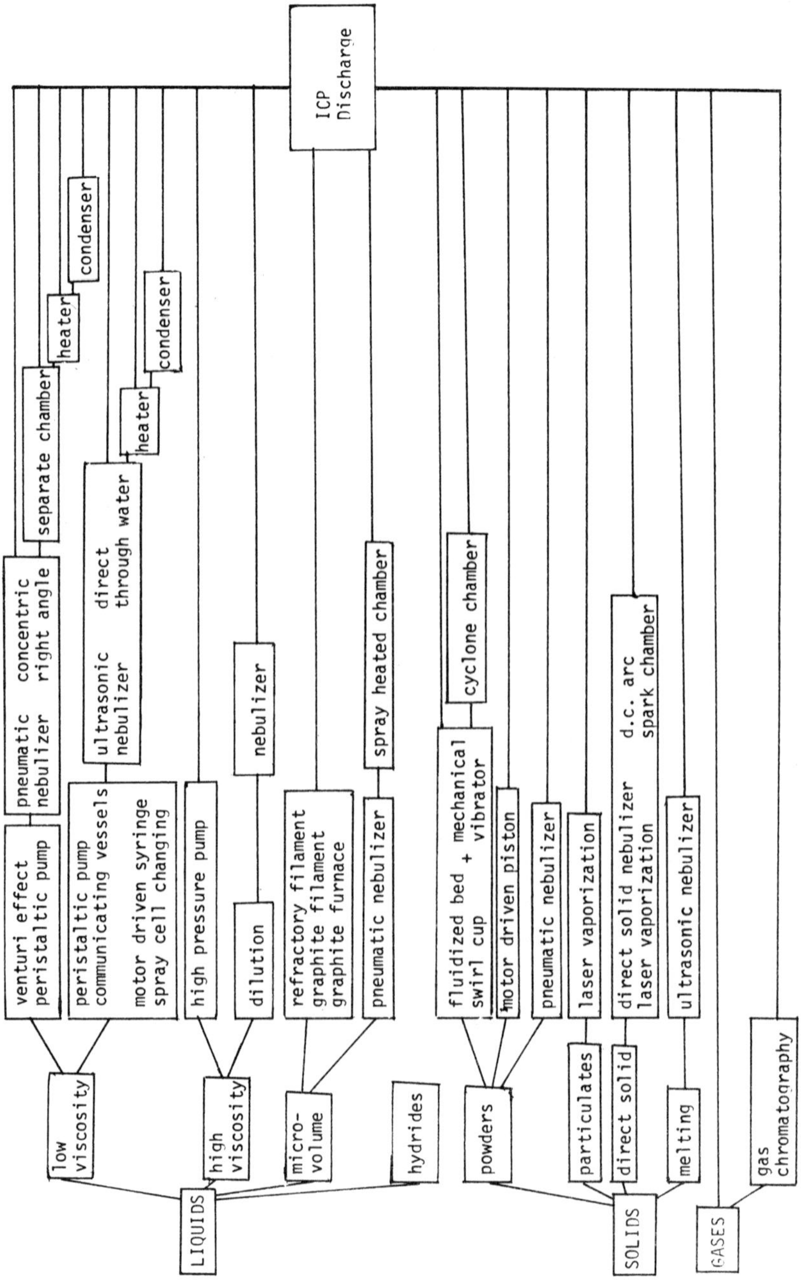

FIGURE II. SAMPLE INTRODUCTION FORMS

one of a variety of nebulizer designs. The solution is sucked up in pneumatic nebulizers, broken into fine droplets, and passed into a spray chamber where the largest droplets comprising the bulk of the solution (approximately 96%-99%) condense and drain away. The remaining fine mist is carried into the discharge by the argon aerosol carrier gas stream.

A complete ICP-AES instrument, represented in Figure 3, consists of a radio-frequency generator, a means of transmitting and coupling the high frequency signal from the generator to the induction coil, the plasma torch with its gas supplies and flow controllers, the sample introduction apparatus (usually a nebulizer and spray chamber), imaging optics, a monochromator or multichannel spectrometer, readout and control system, and information storage and display. A number of the most popular ICP generator systems are listed in Table II. The most widely employed is the crystal-controlled generator operating at 27 MHz with a forward output power range of 0.6 to 3 kW. Essentially identical discharges can also be generated by other circuit designs and at other high frequencies in the ranges listed in Table II. A substantial difference occurs when nitrogen is introduced as the plasma gas, because the ICP must operate with a significantly higher input power(4).

In a multielement ICP-AES system, the spectroscopic

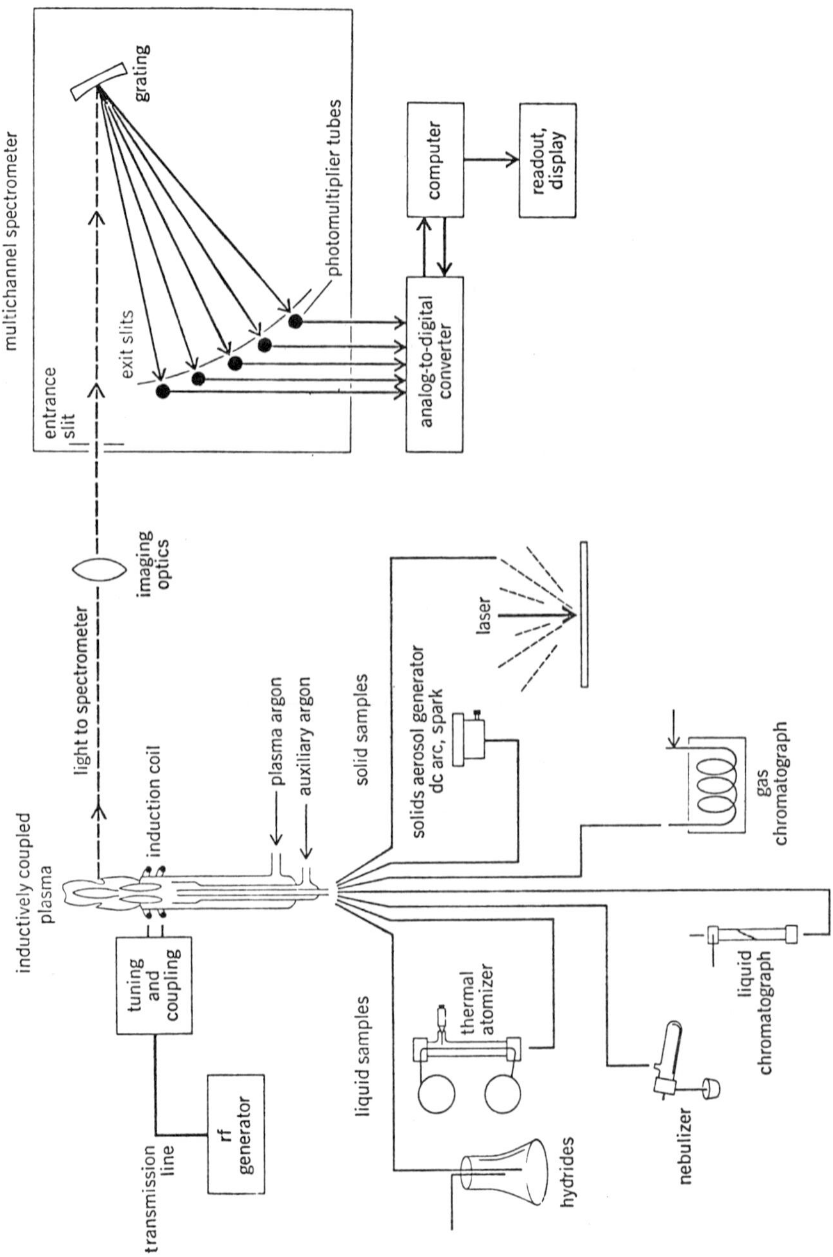

FIGURE III. ICP-AES INSTRUMENT (6)

TABLE II. CHARACTERISTICS OF TYPICAL ICP SYSTEMS

CHARACTERISTICS OF TYPICAL ICP SYSTEMS

Generator	Frequency MHz	Power kW	Gases
Crystal-Controlled	27, 41	1-5	Ar/Ar
Free-Running	4	4-7	N_2/Ar/Ar
	27, 50	1-2	Ar/Ar
Tuned-Line	40	0.6-6	Ar/Ar
Tuned-Cavity	27	1-10	N_2/Ar/Ar

instrument may be one of the types listed in Table III. The spectrograph, especially when combined with a computerized microphotometer, provides full multielement detection capabilities and background correction at moderate costs in exchange for the inconvenience and time required to process the photographic emulsion. Multichannel spectrometers are the most commonly applied instruments for routine ICP multielement analysis and yield simultaneous, almost instantaneous results with computer-controlled systems. The manual scanning monochromator provides versatility for single-element analysis and is a moderate cost multielement sequential analyzer under computer program control. Although echelle spectrometers have been used in research

TABLE III. MULTIELEMENT ICP-AES SYSTEMS

Spectrograph

- Manual Densitometer
- Computerized Densitometer

Multichannel Spectrometer

- Multiple Exit Slits & PMT's
- Linear Photodiode Array

Scanning Monochromator

- Manual, Analog or Digital
- Computer Programmable
- Photodiode Array

Echelle Spectrometer

- Single Channel
- Multiple Exit Slits & PMT's
- SEC Vidicon Camera
- Image Dissector PMT

Interferometer

laboratories with the ICP source, multichannel photomultiplier, vidicon, or image dissector camera systems used in conjunction with the echelle spectrometer are currently under commercial development for multielement analysis.

Operating Conditions

In general only three experimental parameters need be routinely altered to reach optimum conditions for simultaneous multielement analysis with good sensitivity and minimum influence from the sample environment. Although the optimum conditions depend upon the particular equipment arrangement, for a fixed set of conditions, the three analytically adjustable parameters include the input power from the high frequency generator, the aerosol carrier gas flow rate, and the zone of observation above the top of the induction coil.

According to Boumans and co-workers (8), a generalized set of rules can be outlined for ICP-AES under fixed ICP compromise operating conditions for simultaneous multielement analysis indicated in Table IV. For

TABLE IV. GENERALIZED RULES FOR ICP-AES ANALYSES

(1) Select the (most) sensitive analysis line(s) and determine approximate detection limit(s).

(2) Check with "aqueous" solutions for spectral interferences from expected concomitants. Check for blanks.

(3) Compute dilution factor so that analyte concentration in sample solution will be greater than or equal to 100xDL.

(4) Construct analytical calibration curve with "aqueous" reference samples around the analyte concentration and match the acid concentration of references to that of eventual sample solution.

(5) Check absence of matrix effects and adapt calibration curve when necessary.

(6) Dilute sample and measure.

single-element analysis, the most sensitive analysis line is selected, detection limits checked, and spectral interferences and blank values evaluated. In multielement analysis employing a commercial multichannel spectrometer, the spectral lines have been selected by the manufacturer, and the evaluation of spectral line interferences and solution blanks becomes foremost for the sample under consideration. Boumans et al (9) also presented guidelines for the selection of analytical lines summarized in Table V. In the ICP discharge, many elements show the greatest sensitivity when their ion lines rather than their atom lines are employed. Standard wavelength tables are useful when used only for order-of-magnitude comparisons within a level of ionization, and intensities of ion lines are about two

TABLE V. GUIDELINES IN SELECTION OF ANALYSIS LINES

1 Use the most sensitive line(s) and look for less sensitive line(s) in case of spectral interference.

2 Select the lines according to
  a) experience with an ICP monochromator system
  b) general data on ICP spectra in the literature
  c) NBS Tables of Spectral-Line Intensities (NBS copper arc):
    - tabulated intensities are, as to their order of magnitude, representative for the relative detection limits within <u>either</u> the ionic <u>or</u> the atomic spectra,
    - ionic lines are about two orders of magnitude more sensitive than atomic lines than would follow from the Tables.

orders-of-magnitude more sensitive than atom lines from the standard tables. Experimental studies indicate that the emission intensities of ionic lines (e.g., Ca II) are more dependent upon both power level and aerosol carrier gas flow than atom emission (e.g., Ca I), but the spatial distribution of emission from ions axially above the discharge is not altered as significantly as that of atom lines (10). Other factors important in spectral line selection include freedom from interference, favorable background characteristics, minimal stray light effects, linearity of analytical curve, and, in a multichannel spectrometer, the physical space with which exit slits, mirrors, and photomultiplier tubes can be positioned (11).

The excellence of spectroanalytical results obtained by ICP-AES stems from the ICP's high stability, low noise and background, and freedom from interferences when operated under appropriate experimental conditions. Because of the unique spatial distributions of analyte emission and background found in the ICP discharge, the spectral background intensities are approximately 100 times smaller and three times more stable than with other emission sources(12). When compared to the capacitively coupled microwave discharge, for example, not only is the noise of the background lower but it remains unchanged as large amounts of an easily ionizable element (Cs) are added, in contrast to the increase for the microwave discharge(13). In practical operation, however, sufficient warm-up time is required after the instrument is turned on in order for it to stabilize and reach a thermal equilibrium. This takes 15 to 30 minutes. Once the power supply and components in the discharge source have reached a stable operating condition, the long term (i.e., 1 day to 0.5 year) stability is only slightly poorer (2-5%) than the short-term stability (i.e., 0.5 to 8 hours) which is normally less than ±1%. The frequency of standardization during routine operation affects the analytical results, in general, by correcting for instrumental system drift. As indicated in Table VI, at 100 times or more above the limit of detection, the precision is of the order of 1-3% when calibration is repeated every 30 minutes. If standardization is repeated at

TABLE VI. STANDARDIZATION FREQUENCY EFFECT ON PRECISION

	Concentration Multiple Above Detection Limit			
Frequency hours	10-100	100-10^3	10^3-10^4	10^4-10^5
0.5	3-7%	1-3%	1-2%	1.5-2%
2	5-10%	2-6%	1.5-2.5%	2-3%
8	8-15%	3-10%	3-7%	4-8%

M.D. Silvester & R.B. Cruz, Barringer Research, March 1976

8-hour intervals, then the precision at all concentration levels is degraded by a factor of two to four. In routine analysis, the precision can be related to the concentration multiple of the detection limit and the background equivalent concentration (12). In attempting to isolate some of the instrumental factors affecting precision and accuracy, Watters and Norris reported in an accompanying article data for each stage of the overall ICP analysis. Although detection electronics and plasma emission contributed less than 0.1% and 0.3% respectively, introduction of blank and sample solutions increased the variability to between 0.5 and 1.5% under optimum circumstances. ICP users generally agree that the major source of imprecision is the sample introduction and, in particular, the solution nebulizer system.

The stability of ICP generators, spectrometers, and readout systems are sufficiently reliable that the analytical precision, the limits of detection, and the accuracy depend more upon the type of nebulizer and nebulization characteristics than upon the remainder of the measurement system. Since the limit of detection depends directly upon the rate of solvent-free aerosol reaching the discharge in low-power discharges, ultrasonic nebulizers provide better limits of detection than pneumatic nebulizers (14). This results because ultrasonic nebulizers produce aerosols of greater number density and of more uniform particle size than pneumatic nebulizers. Olson et al (15) recently described a new ultrasonic nebulizer, which is free from chemical attack, easily used with solution samples, and commercially available. After a comparison of ultrasonic and pneumatic nebulizers with and without desolvation, Olson et al. found an order-of-magnitude better detection limit with the ultrasonic nebulizer corresponding to approximately an order-of-magnitude increase in nebulization efficiency. Other considerations enter the picture, however, when practical analyses are investigated. For example, Boumans and de Boer (16) observed a fivefold increase in sample injection rate and limits of detection upon changing from pneumatic to ultrasonic nebulization, but a tenfold reduction in the concentration of alkali salt which would produce a level of interference of less than 10%. Therefore, detection limits with respect to dissolved solid were the same or slightly better with a pneumatic nebulizer because a tenfold higher concentration of dissolved material could be tolerated. A further complication related to nebulizers is the effect of the change of physical property of the solution upon the nebulization and transport efficiency of the sample to the discharge(17-19). The emission intensities drop as the acid concentration of the sample increases owing to the increased viscosity of the acid solutions. In practical analyses, the matching of acid strength between samples and standards is a simple and standard procedure to guarantee accurate results. Thus, accuracy in ICP-AES can be achieved with aqueous sample by using relatively simple reference standards which contain suitable acid concentrations and approximate concentrations of elements which cause spectral or stray light interferences.

The final item of interest in considering the ICP is the cost. Capital equipment costs for a multichannel spectrometer fully equipped with ICP source, computer readout, and at least 20 channels vary with manufacturer from about $1800 to about $3650 per channel without consideration for expenses for preparing the physical location of the system. Operating costs are compared to routine AAS in Table VII (20). On a per element basis, the ICP is less expensive to operate than AAS when nine or more elements are determined in the same sample solution.

TABLE VII. RELATIVE OPERATING COST COMPARISONS

	TOTAL	SAMPLES	COST/SAMPLE	ELEMENTS	COST/ELEMENT
ICP	$465	300	$1.55	22	$0.07
AAS	$180	1000	$0.18	1	$0.18

M.D. Silvester & R.B. Cruz, Barringer Research, March 1976

APPLICATIONS

The inductively coupled plasma discharge is finding applications in a number of fields. Although spectrochemical analysis is probably the most rapidly developing, in spectrophysics the unique spectral excitation properties and stability of the discharge provide identification of new spectra and the measurement of transition probabilities. In the engineering field, the induction plasma is applied commercially for processing materials and synthesis, and in plasma diagnostics the discharge should serve as reliable light source.

In spectrochemical analysis, applications fall naturally into the categories of routine, nonroutine, universal, and process control analyses for both single-element and multielement determinations. Whereas routine analysis deals with many samples of identical or similar properties which are best attacked using a multichannel spectrometer, nonroutine samples generally require "one-of-a-kind" treatment and flexible measurement instrumentation. Thus, finding both a monochromator or spectrograph and a multichannel spectrometer attached to the ICP in the problem-solving analytical laboratory is not uncommon. The scanning monochromator serves both as a single-element analyzer or when scanned to individual wavelengths of interest as a sequential multielement analyzer. Totally automatic, on-line multielement ICP-AES systems for continuous or interrupted sample stream monitoring need yet a totally different instrument design approach compared to today's commercially available operator-assisted laboratory instruments.

Applications of ICP-AES come from many disciplines and include many different sample types for which the ICP-AES instrument may be identical but the sample preparation and presentation techniques differ. A variety of firsthand examples will be found in the accompanying papers, and in Table VIII a list is given of some of the types of samples analyzed by ICP-AES as agricultural, biological, chemical, environmental, and geological materials, metals, and energy-related samples.

Biological Materials

The determination of metal and metalloid concentration levels in biological materials is an active development area for ICP-AES applications, because of the growing need for accurate and rapid multielement analysis of biological fluids, tissues, and hair, for example, in nutritional, pathological, and toxicological investigations. Morrison (21) recently compared ICP-AES with several other instrumental multielement spectroscopic techniques for trace analysis of

TABLE VIII. SAMPLES ANALYZED BY ICP-AES

Metals
- Aluminum
- Copper
- Gold
- Ferromanganese
- Ferrochromium
- Matte-leach Residues
- Nickel
- Silicon
- Steel
- Sm-Co Alloy
- Uranium

Biological
- Blood
- Bones
- Hair
- Teeth
- Tissues
- Urine

Geological
- Geothermal Brines
- Drill Cores
- Minerals
- Ores
- Rocks
- Sediments
- Soil
- Sulfide Concentrates
- Zircon Beach Sands

Agricultural
- Animal Wastes
- Beverages
- Edible Oils
- Fertilizers
- Feeds
- Foodstuffs
- Grain
- Meat
- Milk
- Plant Tissues
- Proteins
- Seaweed
- Plant Oils
- Soil Extracts
- Zooplankton

Environmental
- Air Particulates
- Dusts
- Deposits
- Industrial Effluents
- Ground Waters
- Potable Water
- Sewage
- Sewage Sludge
- Saline & Seawater
- Waste Waters

Chemical
- Inorganic Chemicals
- Food Chemicals
- Etching Baths
- Mineral Acids
- Organic Solvents
- Organic Chemicals
- Plant Process Solutions
- Plating Solutions
- Polymers
- Resins
- Rare Earth Compounds
- Semiconductors

Energy
- Coal
- Coal Ash
- Fly Ash
- Liquefaction Product
- Oil Shale
- Naptha
- Gasoline
- Petroleum
- Shale Oil
- Residual Fuels
- Tar Sands
- Used Oil

biological materials, based upon the effectiveness of these methods in the analysis of serum and/or whole blood. For literature values for 14 essential and 12 toxic elements in serum, ICP-AES trailed electrothermal atomization atomic absorption spectrometry and spark source mass spectrometry in sensitivity, but surpassed instrumental neutron activation analysis, x-ray fluorescence, and dc arc emission spectrography. Since ETA-AAS is a single-element technique and SSMS requires solid samples previously dehydrated and stripped of organic constituents, ICP-AES holds the most promise for multielement analysis.

In the early 1970s, two techniques were described to demonstrate the feasibility of ICP analysis of whole blood and serum for major and minor elements in microvolume sample sizes (22-24). Butler and Kniseley (23) diluted whole blood 1:10 with HCl and serum 1:2 with distilled water prior to the determination of nine elements using either standard additions or synthetic blood electrolyte standards with small (1 mL) or micro (25 μL) volume solution samples. The high electrolyte concentration in blood produced signal enhancement of Al later identified as the result of unwanted radiation from Mg and Ca.

Of the potential sources of interference in many practical ICP analyses, unwanted radiation must be considered a major one(25). Stray light results from optical defects within the spectrometer and appears as shifts in background and signal levels related in magnitude to the concentration of sensitive elements such as Ca, Mg, and Al in the sample matrix. With appropriate spectrometer design, the effect of stray light can be reduced so that routine computer corrections can be made based upon blank and standard solutions containing approximate concentrations of these electrolytes. In the ICP discharge, Stark line broadening and radiative recombination continua from sample concomitants (e.g., Al, Ca, Mg) produce spectral background interferences which cannot be reduced instrumentally as is done for stray radiation.

Among the other possible interferences in ICP-AES are the effects of the presence of easily ionizable elements in the sample solution and the formation of refractory compounds in the discharge during the atomization step(19,26). The addition of relatively high concentrations of alkali elements does not seriously degrade the performance of the ICP when the proper operating conditions are selected. For simultaneous multielement analysis the effects can be kept to $\pm$10% or less by selecting compromise conditions for ICP operation(14,27). In single-element analysis, the matrix effects can be reduced to negligible levels.

The formation of refractory compounds in the ICP

discharge does not limit the applications of the ICP as it does in other electrical discharges, such as the microwave plasma (13,26) or combustion flames. For example, the presence of phosphate in calcium-containing solutions produces no calcium emission depression over wide calcium-to-phosphate ratios when compromise operating conditions for simultaneous multielement analysis are used (19). At very high ratios, the salt content of the test solution increases the viscosity sufficiently that a decrease in transport efficiency occurs and the emission signal falls. However, when the calcium emission is ratioed to an internal reference element signal, the transport effect can be compensated. However, the precision becomes somewhat poorer. In practical terms, this means that phosphorous in the 0.3 to 1% concentration range can be determined with a precision of $\pm$2% in milk solutions (28).

Dahlquist and Knoll (19) recently evaluated wet and dry digestion sample preparation procedures for ICP-AES analysis of biological tissue samples. Freeze-dried biological tissues were dissolved using a wet digestion procedure with a mixture of nitric and perchloric acids, and in the dry ashing procedures, tissues were ashed at elevated temperatures prior to dissolution in single mineral acids. A practical limit of 1 g of dry tissue per 50 mL was found using the nitric acid-perchloric acid digestion owing to the limited solubility of alkali and ammonium perchlorates. Dry ashing provided concentrated solutions of 1 g dry tissue in 5 to 10 mL final volumes. This is the only significant limitation of the wet-digestion procedure, but at this dilution, a number of biologically significant elements are close to the concentration limit for workable quantitative analysis. Evaluation of the short-term precision for samples prepared by the two techniques indicated that the ash extract aliquots were not homogeneous. A relatively small systematic error or difference in sample preparation was detected statistically for wet-digested samples, although the precision was within the certified range of uncertainty for the standard tissues used in the test. The ensemble results for nine elements in biological materials including human sera using one set of multielement synthetic reference solutions with one set of instrumental operating conditions showed a high degree of correlation of ICP determinations with assigned content, very small bias, and a good mean recovery. For comparison of the ICP results with other multielement instrumental methods, the reported ensemble relative deviations and the overall bias were not significantly different from the uncertainties of standard reference materials.

These analytical results demonstrate that ICP-AES is a rapid and efficient analytical technique, and interferences observed with other spectrochemical sources are either not measurable or are minimal under practical operating conditions with the ICP. The ICP provides precision and accuracy sufficient to examine critically sample preparation procedures for biological as well as other materials on a comprehensive multielement basis. The ICP also provides simultaneous multielement determinations with greater relative sensitivity for a broader range of biologically significant elements than any other multielement technique commercially available.

Universal Spectrographic Analysis

Although many applications of the ICP source for spectrochemical analysis have been performed using photoelectric detection, the ICP also serves equally well as a source for spectrographic analysis. In practical terms, the majority of emission spectroscopy laboratories are equipped with one or more spectrographs which probably have supported for years qualitative and quantitative analysis with arc and spark discharges. These tools need not be discarded upon adoption of the ICP source into the spectrographic laboratory. With relatively direct modifications, the ICP discharge can be adapted to almost any spectrograph. When spectrographic ICP emission analyses are performed, the general experience has been favorable with precision values exceeding, at times, the best obtained previously with arc and spark sources.

Allemand and Benzie (29) described a commercially available ICP stand which fits an optical bar attached to a 3.4-m Ebert spectrograph. With this arrangement, both the arc-spark stand and the ICP stand could be used without readjustments.

Recently Boumans and co-workers (9) described their novel approach to the universal spectrographic analysis of solutions with an ICP discharge. This strategy takes advantage of all the unique properties of the ICP discharge and the speed and operational simplicity of a computerized microphotometer (Table III) for automatic measurement and analysis of photographically recorded spectra. Replacing Addink's universal dc arc method based upon powder samples, the new Philips quantitative survey approach uses solution samples with the ICP and meets the increasingly severe demands for precision and accuracy being placed upon the now-classical dc arc universal method. In order to match the limited dynamic range of the photographic emulsion with the large dynamic range of the ICP source, Boumans et al devised a scheme, represented in Table IX, of multiple exposures so that the background and gross line signals of all major and minor constituents of interest exhibited a photographic transmittance in the optimum range between 5

TABLE IX. PHILIPS QUANTITATIVE SPECTROGRAPHIC SURVEY ANALYSIS

Exposure	Spectrum	Sample Conc. (% w/v)	Filter T-Step	Exposure Time
1	a	0.01	1.0	t
	c		0.1	
2	a	0.01	1.0	0.1 t
	c		0.1	
3	a	0.001	1.0	t
	c		0.1	
4	a	0.001	1.0	0.1 t
	c		0.1	

and 80%. In the conventional dc arc arrangement, intensities were controlled by recording emission through a stepped filter (1:1/3:1/10:1). With the ICP, the steady-state sample introduction with a pneumatic nebulizer and the high temporal stability permits stepped exposure times of t and t/10. Also, the high powers of detection, large linear range, and low matrix effects of the ICP allow dilutions of the solution concentration in 1:1 and 1:10 steps. Thus four separate exposures, two each with two sample concentrations through a two- or three-step filter for the two exposure times, are necessary to cover the large dynamic range. The resulting recorded spectra are processed automatically on a computerized microphotometer using a program which selects from the eight spectra the transmission data for lines and background and then converts the %T values into relative intensity values while taking into account the various attenuation factors. The concentration range covered in the solid sample depends upon the limit of detection for particular elements. Since the majority of elements fall into the detection limit range of 0.1 to 10 ng/mL and with a sample concentration in solution of 0.01 w/v%, the range from 1 to 10,000 ppm for good elements and 100 to 100,000 ppm for poor elements can be covered with the most sensitive lines. For concentration in the 10-100% range, less sensitive lines must be substituted. In the application of this universal quantitative survey analysis approach, concentration calibrations are made using aqueous solution standards in 0.05 M HCl, compromise ICP operating conditions for multielement analysis, and library of spectral interferences based upon actual measurements of the spectra using aqueous solution containing possible interferents. The systematic application of the ICP and a conventional spectrograph provides the flexibility unachievable with multichannel spectrometer to satisfy the need for nonroutine analysis. Although the possibility exists that a photoelectric spectrometer based upon vidicon or image dissector detectors and echelle spectrometers (30) or computerized programmable monochromator (31) may also afford universal quantitative survey analysis with an ICP source, the ICP-spectrograph-computerized microphotometer combination appears to be well along its way to becoming a basic workhorse for nonroutine sample analysis in the modern spectrographic laboratory.

Nonroutine Sample Analysis

Nonroutine sample analysis problems can be classified into three groups according to Boumans et al (8). The first category includes those problems which could be attacked equally well with both ICP-AES and another method, while in the second classification the problems are better solved by ICP-AES because of the direct sample handling and minimized sample preparation required. In the third category, the ICP alone can solve the problem because of its sensitivity, speed, or reliability as in the determination of uranium in rocks (32). As an example of the first classification, Boumans et al (8) demonstrated the determination of the major components Sm and Co in a binary Sm-Co alloy and compared their results with classical chemical techniques. No problem existed in locating a suitable Sm line that was relatively free of Co interference, but only the Co II lines were sufficiently free of Sm interference to be used. When compared to potentiometric and EDTA titrations, the ICP results compared well although the precision was somewhat poorer. However, the confidence interval of the ICP-AES analysis might be improved by carrying out more determinations on a single sample, since only a small amount of sample is required.

In other nonroutine sample analysis, Watson and Russell (33-35) applied ICP spectrographic analysis in the determination of minor and trace elements in chromite ore and ferrochromium, in the determination of Fe in zircon beach sands, in the analysis of matte-leach residues and in copper, lead, and zinc sulfide concentrates. These examples clearly illustrate the second category in which the ICP analysis procedure simplified sample preparation by reducing the need for solvent extractions, preconcentrations, and masking agents commonly found in atomic and molecular spectrophotometric procedures as a consequence of the high sensitivity and large dynamic range of the ICP. To simplify their procedures, Watson and Russell standardized their dissolution technique based upon a fusion with sodium peroxide in a vitreous-graphite crucible with the resulting melt dissolved in acid and a scandium internal reference added. With a 40-second exposure using a 3.4-m Ebert spectrograph, the ICP source was operated at high input power with nitrogen coolant. Analysis of a variety of samples is possible with one set of calibration solutions with this standardized sample preparation technique, because samples and standards contain equal sodium and acid concentrations. No significant matrix effects were observed, and dilution of sample after dissolution was required only when the measuring equipment could not accommodate the wide dynamic range of the emission from the ICP source. Results for the analysis of chromite ore and ferrochromium agree well with chemical analysis. A similar procedure was developed for dissolution of copper, lead, and zinc sulfide concentrates produced in a pilot plant beneficiating a sulfide ore. The sample was fused with sodium peroxide with a small addition of sodium carbonate, and the melt was leached with water and nitric acid. In this procedure the final sodium content reached about 4% and the total solids content was about 7%. In the determination of iron in zircon sands, an alternative flux was tested because a high Fe blank was found in sodium peroxide, but the results were in good agreement with AAS and XRF. When zirconium oxychloride was added to the solutions to determine the influence of zirconium concentration, a 6% decrease in sensitivity resulted and corresponded exactly to the increased viscosity of solutions. Overall precision was on the order of $\pm$2-3%.

In the spectrographic analysis of matte-leach residues, which contained up to 50 percent noble metals along with Cu, Ni, Fe, and S, the sample was again fused with sodium peroxide and roasted at 700°C. The HCl-leached melt provided excellent agreement for the determination of noble metals by AAS although no data was provided for comparison for the base metals. Scott et al (36) have also determined simultaneously platinum group metals and gold in ores using ICP-AES with various sample treatment depending upon the ore grade. For high grade ores, the samples are dissolved in aqua regia, whereas for low grade ores, the noble metals are collected in a nickel sulfide button after fusion in sodium carbonate-sodium tetraborate flux.

In estimating the cost reduction resulting for the employment of ICP-AES, Watson and Russell fixed a value of $58,000 per year for a throughput of 20,000 determinations (34). Their laboratory is also equipped with a multichannel spectrometer-ICP instrument used for the determination of Cu, Zn, and Fe in sulfide ores and the analysis of plant process solutions to name a few types of samples. These solutions contained a wide variety of elements at both high and low concentration levels, and in the past they were analyzed by AAS or spectrophotometry, which often required a closely matched set of standards and separation procedures. In contrast, ICP-AES provides a minimum sample preparation time and simultaneous multielement analysis.

A further example of the simplicity of ICP-AES analysis compared to AAS and other techniques is the determination of nickel at the microgram-per-milliliter level in a 7N sulfuric acid etching liquid containing 10% chromium (8). This analysis places a severe demand on finding an interference-free analysis line because of the high Cr content. Using a scanning monochromator, the majority of sensitive nickel lines were eliminated, and the Ni 341.4 nm line selected although two nearby chromium lines contributed to the background level. The net intensity of the Ni line was found to be virtually independent of the chromium concentration when the background was subtracted. We should expect, however, somewhat poorer precision than if no background interference were present. The analysis results indicated that aqueous standard containing chromium compared favorably to values obtained by standard additions.

A further illustration of a problem solved with ICP-AES otherwise requiring time-consuming separation or preconcentration when attacked by AAS is the determination of tin in 18% sulfuric acid solution obtained from the acid digestion of an organic compound. The analysis required handling a batch of a hundred samples at concentrations of 20 ng Sn/ml. Although electrothermal atomization atomic absorption could perform the analysis at higher concentration ranges, the ICP covered the entire range of interest. Since the sulfuric acid concentration varied considerably from sample to sample, changes in background were determined using a scanning monochromator, and net tin intensities were found to be independent of the background variations.

These examples illustrate the values of both photographic recording and scanning monochromators in line selection and background correction for single-element ICP analysis, and they emphasize the need to provide background correction with multichannel spectrometers. The photographic emulsion naturally allows background correction for all recorded lines, and a spectrograph should be evaluated whenever multielement determinations may be limited by background changes in spectrometric measurements.

Metals

Many types of metals have been analyzed spectrometrically and spectrographically with the ICP discharge. For example, Newland and Mostyn (37) determined Co (90.25-1.25%), Fe (0.2-0.4%), Mn (0.5-0.25%), Ta (0.03-0.35%), Ti (0.8-1.5%), and Zr (0.04-0.2%) in nickel alloys with a nickel internal reference and a two-minute exposure using a large quartz spectrograph. For complete sample dissolution, a four-acid mixture including HF was employed followed by boric acid addition to complex excess HF and prevent attack on the glassware in the torch-injector system. No matrix effects were observed, and the precision was on the order of $\pm 2\%$, which is extremely good for a photographic method.

Newland (38) also demonstrated recently the

ICP-spectrographic determination of traces of Ce (0.02-1.3%), La (0.01-0.5%), Nd (0.01-0.4%), and Pr (0.003-0.14%) in plain carbon and low alloy steels employing an iron internal reference at a precision level of 4 to 6%. Because the iron content of the steels varied, Newland extracted Fe (III) into butyl acetate and added a fixed concentration of iron to the iron-free solutions. Exposures of 2 minutes with an ultrasonic nebulizer and a large quartz spectrograph provided excellent sensitivity for the rare earth elements.

Somewhat earlier Dickinson (39), Kniseley (24), and Butler et al (40) determined spectrometrically alloying and impurity elements in low and high alloy steels, applying both a scanning monochromator, multichannel spectrometer, and a scanning spectrometer. Determination of alloying and impurity constituents in steels of widely varying composition requires consideration of the possibilities of spectral interferences and interelement effects. Selection of suitable analysis lines can be based upon wavelength scans obtained with a monochromator, for example, and with solutions containing approximately 0.5% iron and the individual elements. The refractory metals (Nb, W, Ta, and Zr) present some difficulty in analysis by atomic absorption spectrometry, and rare-earth-element (Ce, La, and Pr) determinations often require prior separations by other techniques. These along with speciality elements such as lead and arsenic at trace levels were determined directly in dissolved steels. Curves in Figure 4 illustrate the linearity of calibration, the dynamic range of concentration and readout, and the long-term precision. During the one-year data collection period, the precision of the slopes of these curves remained less than $\pm$4%. With the calibration linearity, both major and trace constituents in a sample could be determined at a single dilution level. In contrast, the analytical curves obtained by flame atomic absorption and emission are generally linear only over 1.5 to 2 orders of magnitude (41).

Element interferences were found to occur at negligible levels in ICP-AES using 0.5% Fe based solution standards, and all ICP determined concentrations fall within the NBS-certified range although the iron content varies from 99 to 1 wt %. No significant deterioration of detection limits was observed when pure aqueous solutions were compared to those containing 0.5% Fe within the standard factor of 2-3 inherent in the determination of limits of detection. Congruent analytical curves obtained in a variety of matrices illustrate the validity of background correction and absence of measurable interelement effect. As Ohls (42) has noted, the ICP in a steel mill laboratory offers simultaneous analysis of all elements in steel as long as

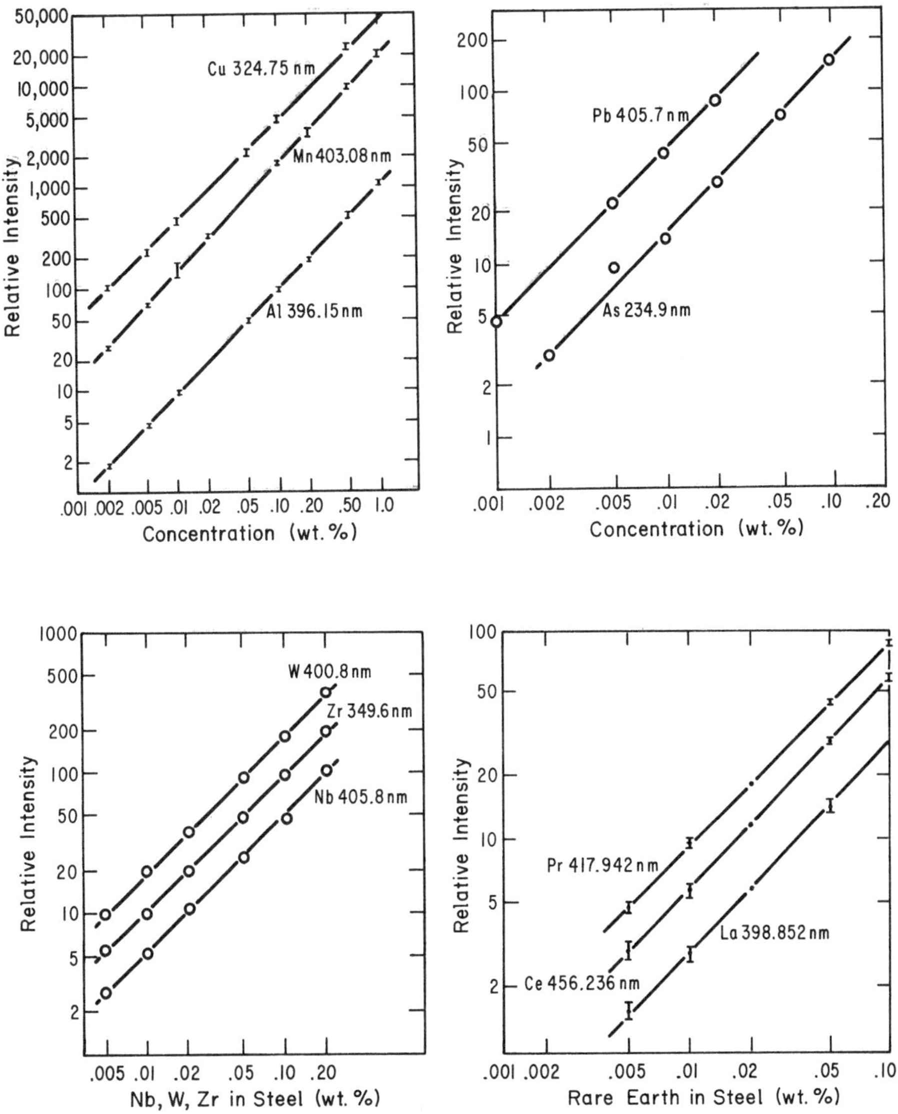

FIGURE IV. CALIBRATION CURVES FOR STEEL SOLUTIONS (24)

they can be put into solution (with the exception of C and S), the easy determination of elements which are difficult to determine by AAS, and the analysis of other materials related to steel production, e.g., refractory oxides, slags, sintered materials, waters, organic solvents, dusts, electrolyte baths, grease, oil, and gasoline.

Oils

Analyses using the ICP discharge are as applicable to metals in organic media as in aqueous media, and a variety of sample types from lubricating, fuel and edible oils to coal liquefaction product, tar sand, and shale oil have been analyzed for metals using ICP-AES.

Fassel et al (43) described the simultaneous determination of 15 different wear metals in lubricating oils using a 2.1-kW discharge and sequential single-channel and multichannel spectrometers. Lubricating oil samples were diluted 1:10 with MIBK and nebulized pneumatically into the discharge at an oil uptake rate of 1 mL/min and a sample aerosol delivery to the discharge of approximately 0.1 mL/min. With a 30-second integration, the analytical cycle time per sample was 1.5 minutes. Introduction of organics into the ICP requires some operating parameter readjustments compared to aqueous solutions to maintain a stable discharge, minimize molecular band emission, and reduce the possibilities of deposits on inner tube surfaces. By increasing the power level, adjusting the torch position with respect to the induction coil, and introducing a small (1 L/min) flow of auxiliary gas, the diluted oil sampled can be analyzed without difficulty. In general terms, the nebulization process depends in a complicated way on the solvent surface tension, density, and viscosity according to Greenfield et al (17, 18), so that caution must always be exercised to accommodate changes in the delivery rate of sample aerosol to the discharge traceable to the physical properties of the organic solvent. Dilution of lubricating oils 1:10 with MIBK appears to reduce the influence of viscosity differences, and analytical curves for nine of 28 elements determined in lubricating oils are linear over 2.5 to 4.5 orders of magnitude (Figure 5) (44). For a 10-second integration, limits of detection before dilution measured with a 0.5-m monochromator at three times the standard deviation of the background signal are given in Table X. Since after dilution low-viscosity jet engine oil and high-viscosity reciprocating engine oil showed negligible difference in viscosity, a single set of reference solutions was used in their analysis by Fassel et al (43).

According to Peterson (44) used lubricating oils may contain suspended microscopic or submicroscopic particles

TABLE X. DETECTION LIMITS OF ELEMENTS IN OIL

Element	Wavelength (nm)	Detection Limit (μg/g of oil)
Ag	328.07	0.02
Al	308.22	0.06
As	234.98[a]	2.
B	249.68	0.2
Ba	233.53	0.03
Be	313.04	0.004
Bi	223.06	0.4
Ca	393.37	0.0006
Cd	226.50	0.09
Co	238.89	0.06
Cr	283.56	0.03
Cu	324.75	0.009
Fe	259.94[a]	0.06
Hg	253.65	0.3
Mg	279.55	0.01
Mn	257.61	0.006
Mo	386.41	4.
Ni	341.48	0.1
P	255.33	9.
Pb	283.31[a]	0.5
Sb	217.59	0.6
Se	196.03	2.
Si	288.16	0.07
Sn	284.00[a]	0.05
Ti	334.94	0.01
V	309.31	0.05
Y	371.03	0.02
Zn	213.86	0.03

a For these elements, a 0.5-m Ebert monochromator was used. Detection limits for all other elements were determined with a multichannel spectrometer (44).

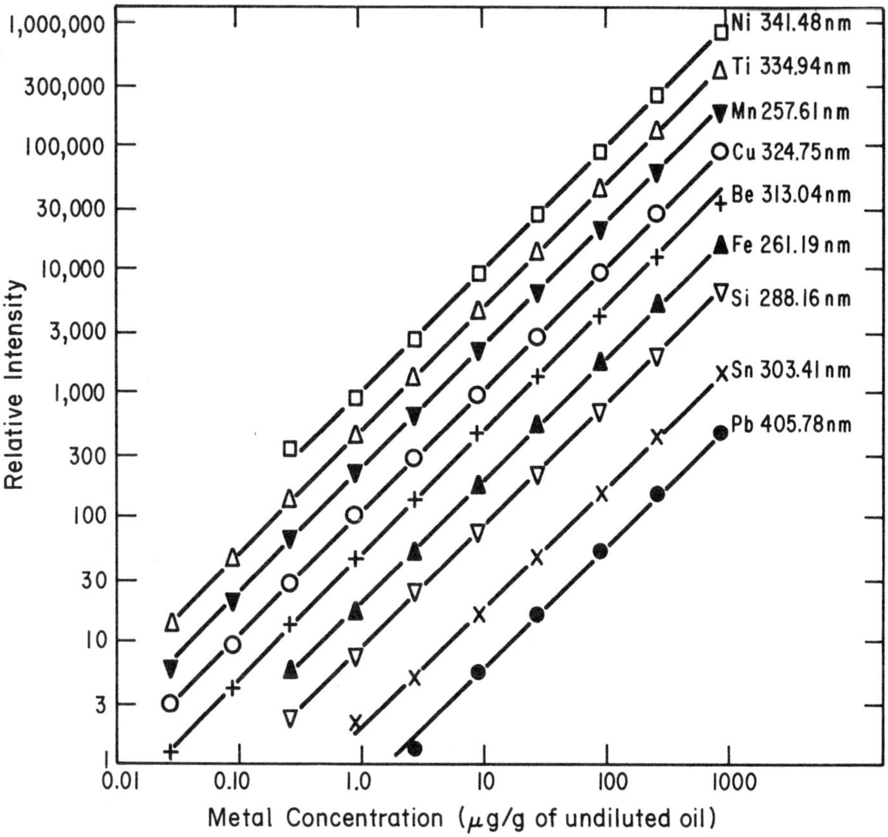

FIGURE V. CALIBRATION CURVES FOR METALS IN OIL (44)

indicative of gradual wear or erosion, or larger macroscopic particles, indicative of rapid failure of oil-wetted parts. Generally, the assumption is made that the atomization source possesses the same efficiency for particles and for dissolved, oil-soluble organometallic compounds from which reference samples are generally prepared. Various techniques are used to minimize the influence of particles in the analysis of lubricating oils, and small particles presented no problem in ICP-AES. However, larger particles do. The noise level observed from the ICP for oil samples containing synthetic iron particle suspensions (0.5 to 4 μm) is greater than that observed for a 10 μg/g organometallic iron solution. Recovery studies for the synthetic suspension gave approximately 10-20% owing to particle settling in the sample container and in the nebulizer sample uptake tubing, low nebulization efficiency and spray chamber loss, and incomplete particle vaporization in the discharge over power ranges of 1.7 to 2.2 kW.

For the analysis of trace metals in fuel oil, toluene or xylene completely dissolved the samples. Detection limits obtained in xylene and toluene in place of MIBK were within a factor of 2 or 3 and 4 respectively. Centrifuged coal liquefaction product samples were dissolved in a 1:1 mixture of MIBK and pyridine at a dilution of 1:10. For the analysis of edible oils, 1-gm samples of crude soybean oil, corn oil, peanut oil, and vegetable oil were diluted 1:4 w/v with MIBK and analyzed for Fe, Mg, and Ca using experimental organometallic reference oil samples. Differences in nebulization characteristics for the corn-oil blank and for lubricating oil blanks resulted in background shifts for several spectral lines, and proper background correction for organic solvents is important.

Volatile organic compounds can also be determined using the ICP without solution nebulization. Denton and Windsor (45) suggested the simultaneous elemental analysis of organic compounds in the ICP when the discharge was attached either to a heated vaporizor or the effluent of a gas chromatograph.

These examples illustrate but a few of the possible applications of ICP-AES. The subsequent seven papers will emphasize other aspects of ICP applications, and as ICP-AES continues to grow the total number of different applications will be limited more by the imagination of the analyst than the fundamental capabilities of the ICP discharge.

ACKNOWLEDGEMENTS

Preparation of this manuscript was supported in part by a grant from the Alcoa Foundation and Department of Energy (Office of Basic Energy Sciences and Division of Biomedical and Environmental Research) contracts.

REFERENCES

1. R.M. Barnes, CRC Crit. Rev. Anal. Chem., 7(3), 203 (1978)

2. P.W.J.M. Boumans, ICP Inform. Newsl., 3, 71, 142, and 173 (1977).

3. V.A. Fassel and R.N. Kniseley, Anal. Chem., 46, 1110A, 1155A (1974).

4. S. Greenfield, H. McD. McGeachin, and P.B. Smith, Talanta, 23, 1 (1976).

5. R.M. Barnes, ed., ICP Information Newsletter, Department of Chemistry, University of Massachusetts, Amherst, Mass. 01003.

6. R.M. Barnes, in "McGraw-Hill Yearbook of Science and Technology," New York: McGraw-Hill Book Co., 1978, pp. 342-345.

7. J.M. Mermet, ICP Inform. Newsl., 2, 70 (1976).

8. P.W.J.M. Boumans, L.C. Bastings, F.J. de Boer, and L.W.J. van Kollenburg, ICP Inform. Newsl., 3, 201 (1977).

9. P.W.J.M. Boumans, A.W. Witmer, F.J. de Boer, and M. Bosveld, ICP Inform. Newsl., 3, 213 (1977); Z. Anal. Chem, 291, 19(1978).

10. T.E. Edmonds and G. Horlick, Appl. Spectrosc., 31, 536 (1977).

11. R.K. Winge and V.A. Fassel, Paper No. 24, Fourth Annual Meeting FACSS, Detroit, 1977; ICP Inform. Newsl., 3, 134(1977).

12. R.M. Ajhar, P.D. Dalager, and A.L. Davison, Amer. Lab.,

8(3), 71 (March 1976).

13. P.W.J.M. Boumans, F.J. de Boer, F.J. Dahmen, H. Hoetzel, and A. Meier, Spectrochim. Acta, 30B, 449 (1975).

14. P.W.J.M. Boumans and F.J. de Boer, Spectrochim. Acta, 31B,355 (1976).

15. K.W. Olson, W.J. Haas, Jr., and V.A. Fassel, Anal. Chem., 49, 632 (1977).

16. P.W.J.M. Boumans and F.J. de Boer, ICP Inform. Newsl., 3, 228 (1977).

17. S. Greenfield, H. McD. McGeachin, and P.B. Smith, Anal. Chim. Acta, 84, 67 (1976).

18. S. Greenfield, H. McD. McGeachin, and F.A. Chambers, ICP Inform. Newsl., 3, 117 (1977).

19. R.L. Dahlquist and J.W. Knoll, Appl. Spectrosc., 32, 1 (1978).

20. M.D. Silvester and R.B. Cruz, ICP Inform. Newsl., 1, 261 (1976).

21. G.H. Morrison, ICP Inform. Newsl., 3, 165 (1977).

22. S. Greenfield and P.B. Smith, Anal. Chim. Acta, 59, 341 (1972).

23. R.N. Kniseley, V.A. Fassel, and C.C. Butler, Clin. Chem., 19, 807 (1973).

24. R.N. Kniseley, Ph.D. Dissertation, Iowa State University, Ames, Iowa, 1973.

25. G.F. Larson, V.A. Fassel, R.K. Winge, and R.N. Kniseley,

Appl. Spectrosc., 30, 384 (1976).

26. G.F. Larson and V.A. Fassel, *Anal. Chem.*, 48, 1161 (1976).

27. D.J. Kalnicky, V.A. Fassel, and R.N. Kniseley, *Appl. Spectrosc.*, 31, 137 (1977).

28. A.M. Gunn, G.F. Kirkbright, and L.N. Opheim, *Anal. Chem.*, 49, 1492 (1977).

29. C. Allemand and G. Benzie, *ICP Inform. Newsl.*, 1, 273 (1976).

30. P.W.J.M. Boumans, *Proc. Anal. Div. Chem. Soc.*, 14, 143 (1977).

31. P.W.J.M. Boumans, G. H. Van Gool, and J.A.J. Jansen, *Analyst*, 101, 585 (1976).

32. R.H. Scott, A. Strasheim, and M.L. Kokot, *Anal. Chim. Acta*, 82, 67 (1976).

33. A.E. Watson and G.M. Russell, *ICP Inform. Newsl.*, 2, 173 (1977).

34. A.E. Watson and G.M. Russell, *ICP Inform. Newsl.*, 3, 273 (1977).

35. A.E. Watson, G.M. Russell, and G. Balaes, *ICP Inform. Newsl.*, 2, 205 (1976).

36. R.B. Wemyss and R.H. Scott, *Anal. Chem.*, 50, 1694 (1978).

37. B.T.N. Newland and R.A. Mostyn, *ICP Inform. Newsl.*, 2, 135 (1976).

38. B.T.N. Newland, ICP Inform. Newsl., 3, 263 (1977).

39. G.W. Dickinson, Ph.D. Dissertation, Iowa State University, Ames, Iowa, 1969.

40. C.C. Butler, R.N. Kniseley, and V.A. Fassel, Anal. Chem., 47, 825 (1975).

41. G.F. Kirkbright and A.F. Ward, Talanta, 21, 1145 (1974).

42. K. Ohls, ICP Inform. Newsl., 2, 33 (1976); 2, 357 (1977).

43. V.A. Fassel, C.A. Peterson, F.N. Abercrombie, and R.N. Kniseley, Anal. Chem., 48, 516 (1976).

44. C.A. Peterson, Ph.D. Dissertation, Iowa State University, Ames, Iowa, 1977.

45. D.L. Windsor and M.B. Denton, Appl. Spectrosc., 32, 366 (1978).

SAMPLING APPROACHES FOR THE INDUCTIVELY COUPLED PLASMA IN PRACTICAL ANALYSIS

Richard F. Browner

INTRODUCTION

The place of RFICPOES in the scheme of trace and micro elemental analysis is shown in Figure 1.* This, of course, refers to determination range only, and does not consider many factors which are often of crucial importance in practical analyses, such as the accuracy of the technique, sample preparation necessary, speed of analysis and other important criteria. On the basis of determination range only, RFICPOES occupies a respectable place as part of the atomic absorption/emission/ fluorescence group of techniques, whilst being somewhat inferior (10^3x approximately) to activation analysis and spark source mass spectrometry, and considerably inferior to radio isotope analysis (10^5x approximately). However, practically, the simplicity of operation of RFICPOES, coupled with its reliability and its multielement capability at moderate cost, give it a considerable advantage in most laboratory

*This somewhat optimistic Figure is based upon the earlier work of Tolg.[1]

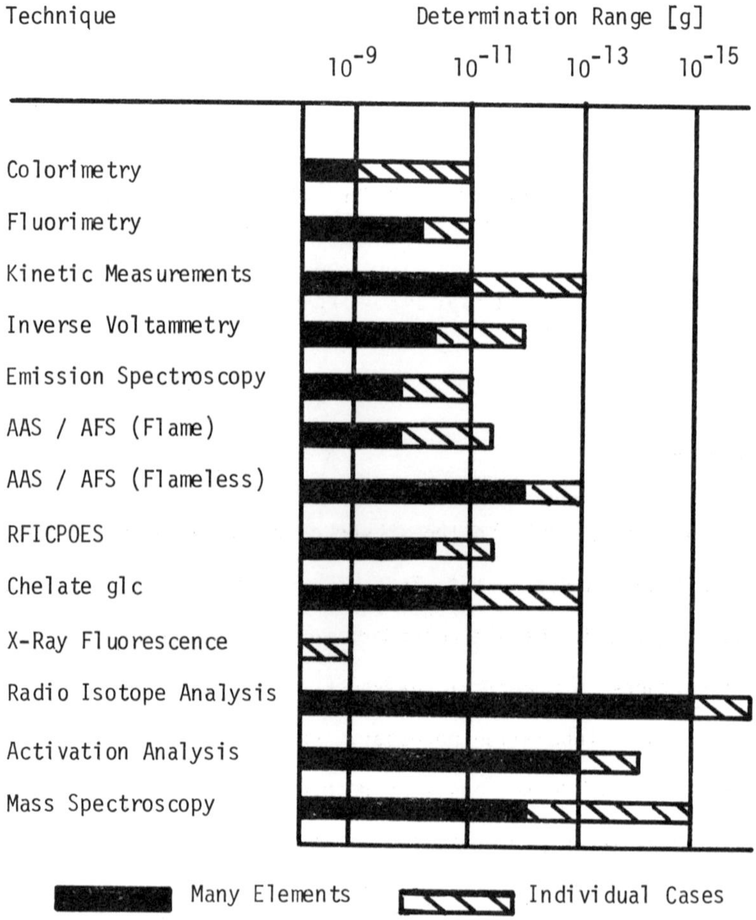

FIGURE I. TECHNIQUE DETERMINATION RANGES

situations. It may be seen that the plasma does not have, at the present time, any great advantage over AAS/AFS in terms of determination range. However, the considerable number of elements with detection limits 5x or more better than for AAS/AFS is a factor favoring the plasma.

One final point of interest which can be seen from Figure 1 is the potential which can be envisioned for the development of the techniques considered. Currently, a major limitation on the performance of RFICPOES (also with flame AAS, but to a lesser degree) is the efficiency with which the sample is transferred from the original material to the plasma itself. Currently, the most usual technique is to dissolve the sample in dilute acid, after the destruction of organic material if present, and then to nebulize pneumatically a fine droplet mist into the plasma. In this process, approximately 80% of the sample is discarded because of its excessively large particle diameter (>10 μm). Clearly, this is an unacceptable situation, and requires the development of alternate means of sample introduction.

SAMPLE INTRODUCTION TECHNIQUES

Some possible routes for sample introduction, starting from either solid, liquid or gaseous materials are shown in Figure 2.

Solid Samples. The prospect of introducing samples directly to a plasma from the original solid form is a very attractive one as it bypasses all the time-consuming and possibly sample-contaminating steps of ashing, acid dissolution or fusion. It is possible to vaporize solid materials directly by arc,[2] spark[3] or laser[4] vaporization techniques and pass the vapor directly into the plasma in a stream of

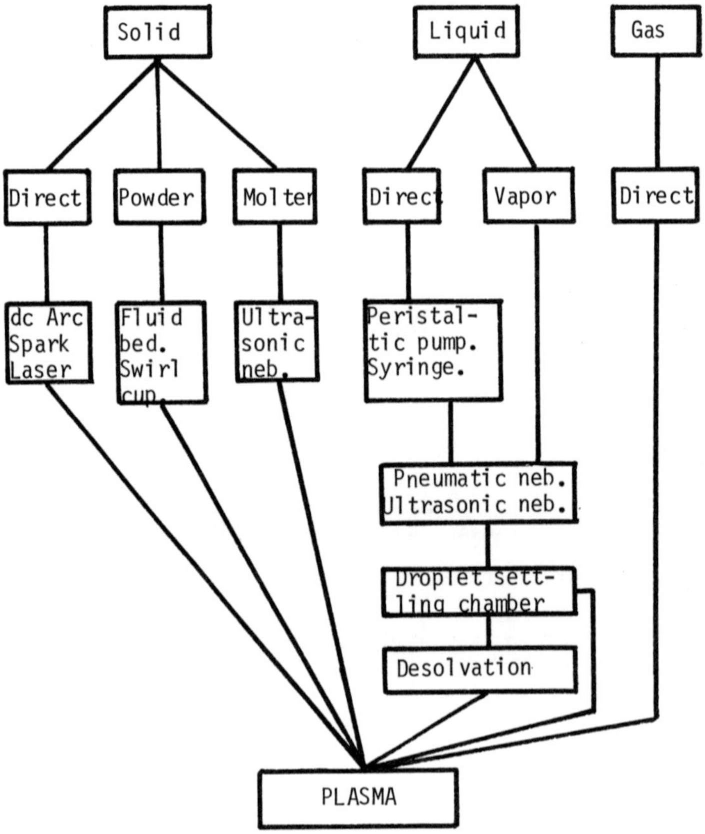

FIGURE II. SAMPLE INTRODUCTION ROUTES FOR RFICPOES

argon. With arc and spark vaporization, the sample must be electrically conducting (which is a limitation on the approach) and the high power capabilities of the CO_2 laser necessary for effective thermal vaporization produces cost limitations and also sampling problems with this approach. Nonetheless, these techniques work well with certain sample types, and short-circuit many tedious stages of preparation.

If the solid is in a finely divided form (10 μm particles or less) it is sometimes possible to introduce a steady stream of particles into the plasma using a fluidized bed[5] or a swirl cup.[6] However, mechanical problems of tube blockage and cross-contamination limit the utility of this approach.

If the solid sample is melted in a furnace (e.g., in a steel mill), the finely divided droplets can be induced by cavitation effects with an ultrasonic horn.[7] The droplets condense in an argon stream and are carried in this up into the plasma.

Liquid Samples. Most typically, as discussed before, the sample is in a liquid form, usually as an acidic aqueous solution. Here, the sample must be converted into droplets ideally <10 μm diameter in order that effective vaporization and excitation can occur. For reproducibility of sample introduction, a peristaltic or syringe pump may be used to propel the sample to the nebulizer. Droplets are produced usually from a pneumatic nebulizer,[8,9] although ultrasonics have been used recently,[10,11,12] and usually the larger droplets are settled out before the aerosol is passed to the plasma. Desolvation can be used, but sample carryover is usually severe in this instance.

Vapor Samples. Volatile hydrides have been used for plasma work, analogously to AAS approaches,[13] but only a few elements can be determined (e.g., As, Se, Sb, Sn, Pb, Bi) and interferences are severe. Hg can also be generated with $NaBH_4$. It is possible, of course, to vaporize the sample directly using an electrothermal device, such as a graphite rod or tantalum boat, and several workers have used this approach.[14,15] However, severe interference problems can be anticipated for samples with complex matrices. Whatever approach is used, for solid, liquid or vapor, the geometry of the RFICP requires that the sample be introduced in a low flow rate argon stream (approximately 1 L min^{-1} flow) into the central hole of the torroidal plasma (Figure 3).

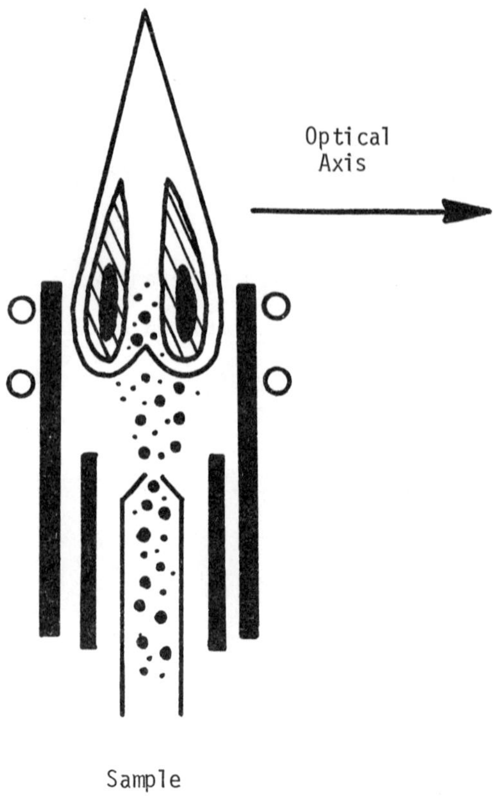

FIGURE III. SAMPLE INTRODUCTION TO RF PLASMA

NEBULIZATION

The very real advantages of using a liquid sample (e.g., ease of handling, homogeneity of sample) have contributed to the continuing wide use of the pneumatic nebulizer, despite the severe restrictions imposed on its geometry by the need for a low flow rate of argon. Typical designs, which both require the use of fine orifices and manufacture to small tolerances, are the Meinhard[8] concentric design and the micro cross-flow design of Kniseley, Amenson, Butler and Fassel[9] (Figure 4). Both of these systems generally work well, but will block

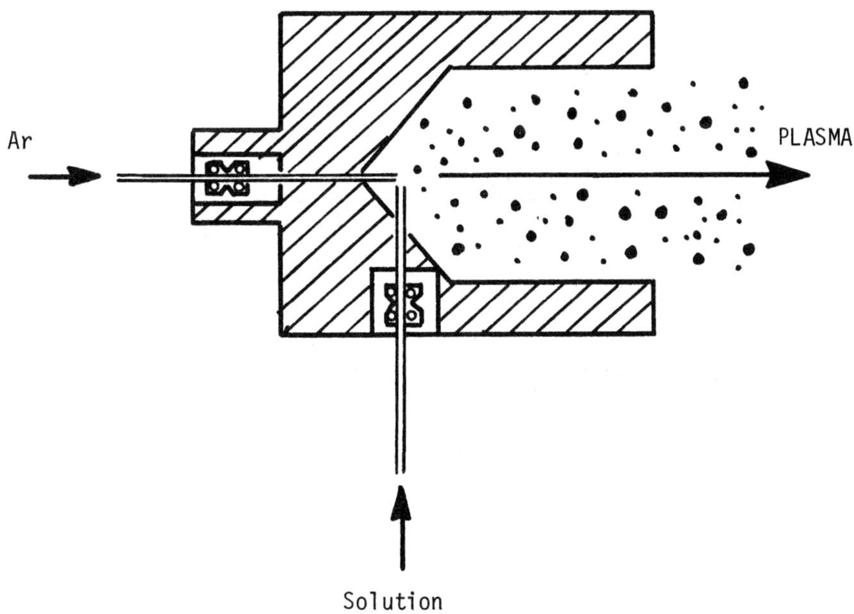

FIGURE IV. MICRO-PNEUMATIC NEBULIZATION

rapidly unless solutions are particle-free and also have low solid content ($\leq$ 4% total dissolved solids).

Ultrasonic nebulization has many attractions for plasma work. The fine droplet spray results from cavitation or geyser formation and so high velocity gas streams are unnecessary and the only gas flow requirement is an adequate flow to remove the droplets formed. The most recent development is by Olson, Haas, and Fassel,[12] based on the earlier work of Kirsten and Bertilson[16] and Stupar and Dawson.[17] Here, a constant loading of the piezoelectric crystal is provided by

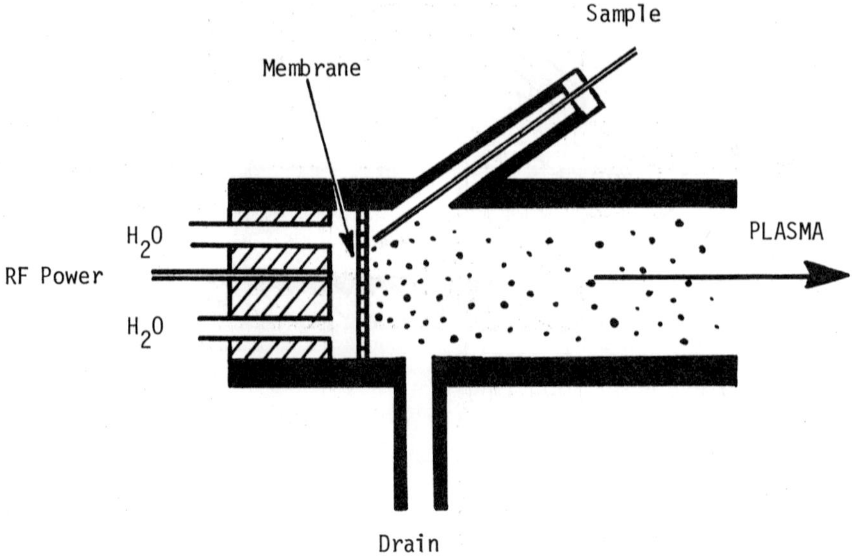

FIGURE V. MICRO-ULTRASONIC NEBULIZATION

using a thin film of liquid which flows from a capillary tube positioned just above the surface of the transmitting membrane (Figure 5).

VAPOR PRODUCTION

The production of volatile species, which can be thermally vaporized into a plasma, is an attractive means of bypassing the usual acid dissolution/nebulization procedure. The work of Skogerboe, Dick, Pavlica and Lichte[18] with metal chlorides is a pointer in this direction. These workers produced volatile chlorides directly by reacting

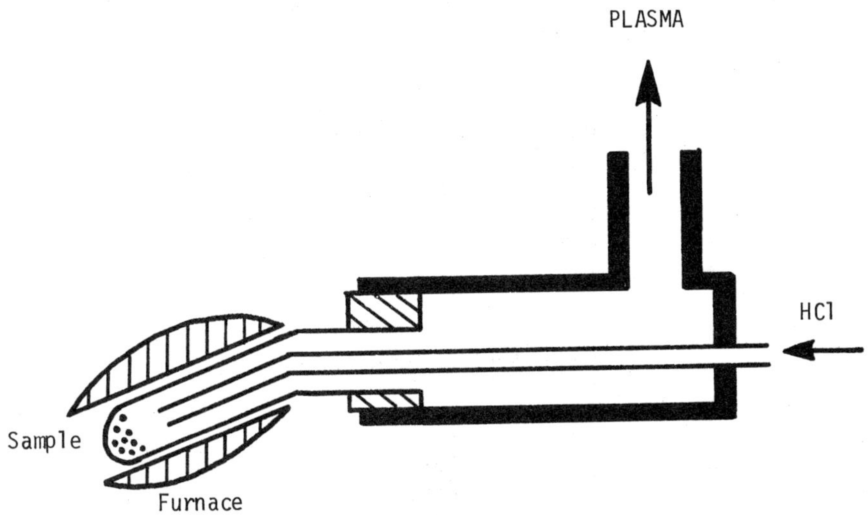

$$M_nY_m\ (s) + Xs\ HCl\ (g) \rightarrow MCl_m\ (g) + H_nY$$

FIGURE VI. SAMPLE INTRODUCTION AS VOLATILE CHLORIDE

gaseous HCl with the sample and then vaporized the chloride from the reaction site by electrothermal means. At furnace temperatures used (up to 1000°C) about 30 elements have sufficiently volatile chlorides to be applicable. Originally this technique was used with flame atomization. We have used the technique successfully in our laboratories, and we are attempting studies on other classes of compounds to which this approach could be applicable (e.g., volatile metal chelates). The prospect of selective volatilization is also very attractive (Figure 6).

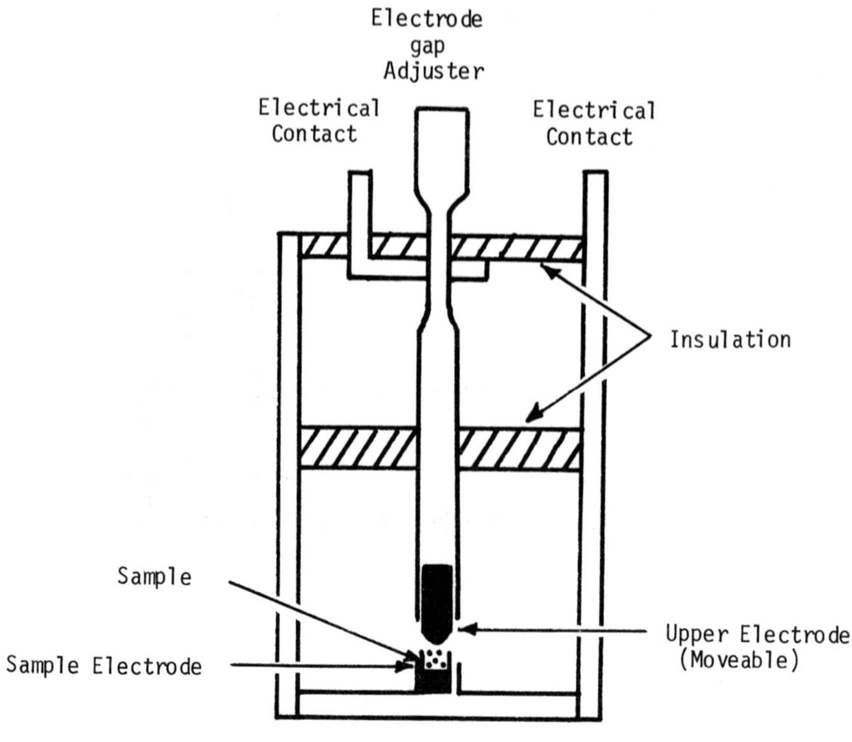

FIGURE VII. CONDENSED SPARK SAMPLE APPARATUS

UNDERWATER SPARK

To return to the problem of sample handling, one of the many difficulties encountered in dealing with real samples is the need to convert these to a form suitable for introduction to the plasma. For many sample types this still necessitates producing a liquid sample, often an aqueous solution, and this can be time-consuming, especially if a considerable amount of organic material must be destroyed. An approach to the production of metallic quasi-suspensions, which behave effectively as metal solutions, was given by Ghiglione, Eljuri and Cuevas.[19] A condensed spark (Figure 7) is formed underwater between two metal electrodes and produces a quasi-suspension of the electrode material. Recently, we have replaced the metal electrodes with graphite electrodes and obtained very promising data with organic samples, by-passing the need for acid digestion.

CONCLUSIONS

In summary, it may be stated that the sample requirements of the RFICP are quite severe, but that further development of the RFICP to a position of full utilization requires a better understanding of the sample introduction step, in order that appropriate techniques may be developed.

REFERENCES

1. G. Tölg, Talanta 21, 327 (1974).

2. R.K. Winge, V.A. Fassel and R.N. Kniseley, Appl. Spectrosc. 25(6), 636 (1971).

3. H.G.C. Human, R.H. Scott, A.R. Oakes and C.D. West, Analyst 101, 265 (1976).

4. F.N. Abercrombie, M.D. Silvester and G. Stoute, 28th Pittsburgh Conference on Analytical Chemistry and Applied Spectroscopy, Paper No. 406 (1977).

5. R.M. Dagnall, D.J. Smith, T.S. West and S. Greenfield, Anal. Chim. Acta 54, 397 (1971).

6. M. Kashiki and S. Oshima, Anal. Chim. Acta 51, 387 (1970).

7. G.W. Dickinson and V.A. Fassel, Anal. Chem. 41(8), 1021 (1969).

8. J.E. Meinhard, ICP Inform. Newsletter 2, 163 (1976).

9. R.N. Kniseley, H. Amenson, C.C. Butler and V.A. Fassel, Appl. Spectrosc. 28, 285 (1974).

10. H.J. Issaq and L.P. Morgenthaler, Anal. Chem. 47, 1661 (1975).

11. P.W.J.M. Boumans and F.J. de Boer, Spectrochim. Acta 30B, 309 (1975).

12. K.W. Olson, W.J. Haas, Jr. and V.A. Fassel, Anal. Chem. 49, 632 (1977).

13. A.E. Smith, Analyst 100, 300 (1975).

14. F.L. Fricke, O. Rose, Jr. and J.A. Caruso, Anal. Chem. 47(12), 2018 (1975).

15. D.E. Nixon, V.A. Fassel and R.N. Kniseley, Anal. Chem. 46(2), 210 (1974).

16. W.J. Kirsten and G.O.B. Bertilsson, Anal. Chem. 38(4), 648 (1966).

17. J. Stupar and J.B. Dawson, Atomic Abs. Newsletter 8(2), 38 (1969).

18. R.K. Skogerboe, D.L. Dick, D.A. Pavlica and F.E. Lichte, Anal. Chem. 47(3), 568 (1975).

19. M. Ghiglione, E. Eljuri and C. Cuevas, Appl. Spectrosc. 30(3), 320 (1976).

FACTORS INFLUENCING PRECISION AND ACCURACY OF ANALYSIS WITH INDUCTIVELY COUPLED PLASMAS

Robert L. Watters, Jr., and John A. Norris

ABSTRACT

In order to analyze Standard Reference Materials for certification, it is necessary to establish which instrumental factors contribute to the errors of the analysis technique. The first step is to attain adequate control of the random errors in the entire measurement process. Once this is achieved it becomes possible to evaluate the magnitude of the error and thereby to properly estimate accuracy.

This paper describes some of the problems encountered in realizing adequate control of the random errors associated with the Inductively Coupled Plasma. The effects of some important instrumental parameters on the precision of the analytical signal will be described. Preliminary analysis data will be presented to demonstrate the possible magnitude of systematic errors which arise from instrumental sources.

INTRODUCTION

In many analytical techniques, the complexity of the instrumental system used for an analysis allows many sources of variability to affect the measurement process. Random and nonrandom behavior are always present which combine to determine the levels of precision and accuracy attainable by a given technique. It is necessary to separate these sources of variability so that their influence on the final result may be evaluated. For the Inductively Coupled Plasma (ICP) technique, the instrumental response is calibrated using solutions of known composition. It is then possible to relate a measured intensity to the concentration of a particular emitting species.

Replicate measurements of standards and samples are made to estimate the true mean concentration for each element of interest. The variance of the technique and resulting confidence intervals are estimated by the statistic s^2. The use of the average and relative standard

TABLE I. TYPICAL ICP OPERATING CONDITIONS

Forward Power	1.0 kW
Reflected Power	4 W
Coolant Flow	20 L/min
Nebulizer Pressure	16 PSIG
Sample Uptake	1.5 mL/min
Observation Height Above Coil	18 mm
Signal Integration Period	30 sec

deviation (RSD) statistics assumes that the errors in the measurement process are random and normally distributed. Since each measurement of a standard or sample consumes a finite amount of time, control of the errors of the process must extend over the entire run sequence. This may be up to an hour or more depending on the extent of replication. It is necessary, therefore, to separate and characterize the sources of variability in the instrumental system over this time period.

INSTRUMENTATION

The plasma is generated by a Plasma-Therm, Inc.* high-frequency supply and matchbox. Automatic power control and tuning are used for all the data presented. The plasma torch is similar in design to the one supplied with the commercial system except that the quartz elements are mounted individually in a plexiglass base. The design of the base assembly is similar to that described by Boumans and de Boer [1]. Both cross-flow and concentric type nebulizers were used. The data as presented here were nearly identical for either nebulizer. Typical operating conditions are summarized in Table 1.

Analytical signal measurements were obtained with a Jarrell-Ash 750 Atomcomp vacuum polychromator and a Jarrell-Ash Model 82-000 monochromator. The two instruments were mounted at 180° with respect to the plasma so that the same observation region could be imaged into both simultaneously.

*In no instance does the identification of commercial products imply recommendation or endorsement by the National Bureau of Standards.

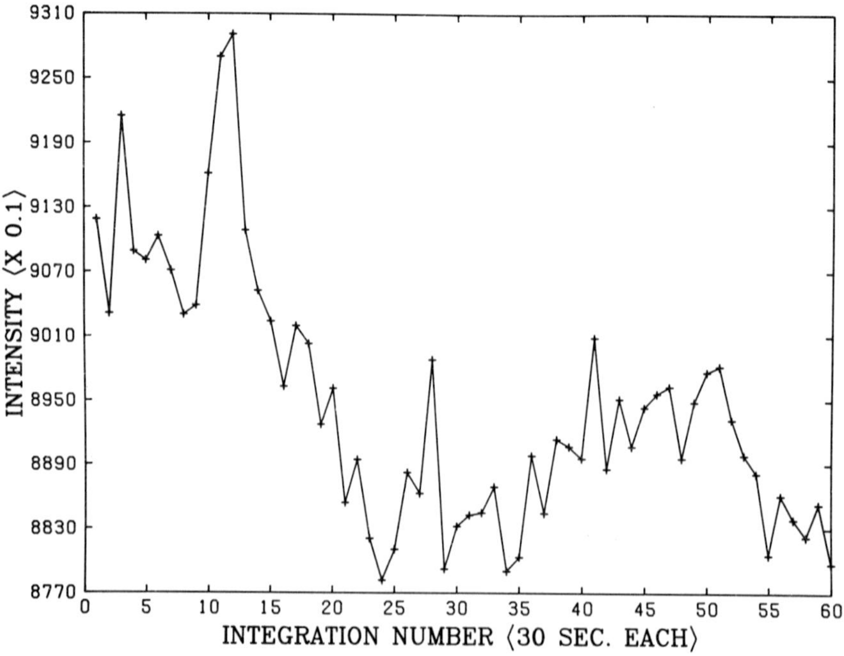

FIGURE I. RUN SEQUENCE PLOT OF INTEGRATED COPPER EMISSION. 327.4nm; aspiration; 1.34% RSD.

RESULTS AND DISCUSSION

Typical total system variability is indicated by the run sequence in Figure 1. The data are plotted for successive thirty-second integrations of photo current from the 327.4 nm copper resonance line. The sample is a mixture of several elements, including copper, all at a concentration of 10 μg/mL. The solution is introduced into the plasma by aspiration with a cross-flow type nebulizer. It is evident that nonrandom variability is present in the system and that systematic error resulting from the range of the data could be more significant than the variance indicated by the RSD. The sources of variability were separated into three groups and examined for their relative contribution to the typical 1 to 2 percent system RSD. The first group includes the photomultipliers, associated electronics, and digital data acquisition facilities of the monochromator and polychromator systems. These were tested using a filament lamp as a light source. Typical levels of variance result in less than 0.1 percent RSD for the same type of run sequence as was shown in Figure 1. This test was repeated with the plasma system running and the spectrometer entrance slit blocked to verify the absence of radio frequency interference from the excitation system.

The second group of parameters tested includes the power generation, coupling, and gas flow facilities of the plasma system. Figure 2 depicts a similar run sequence with an argon line in the plasma being monitored by the monochromator. No sample solution was introduced during this run. While the presence of instrumental drift is apparent, the level of 0.25 percent RSD variability is less than 20 percent of the total system RSD. A similar run measuring the 253.7 nm mercury line during mercury vapor introduction in the sample carrier gas gave a value of approximately 0.5 percent RSD. The temperature dependence of mercury vapor pressure added to the variability in this case. These experiments characterize the base system behavior when the ICP technique is applied to gaseous sample analysis.

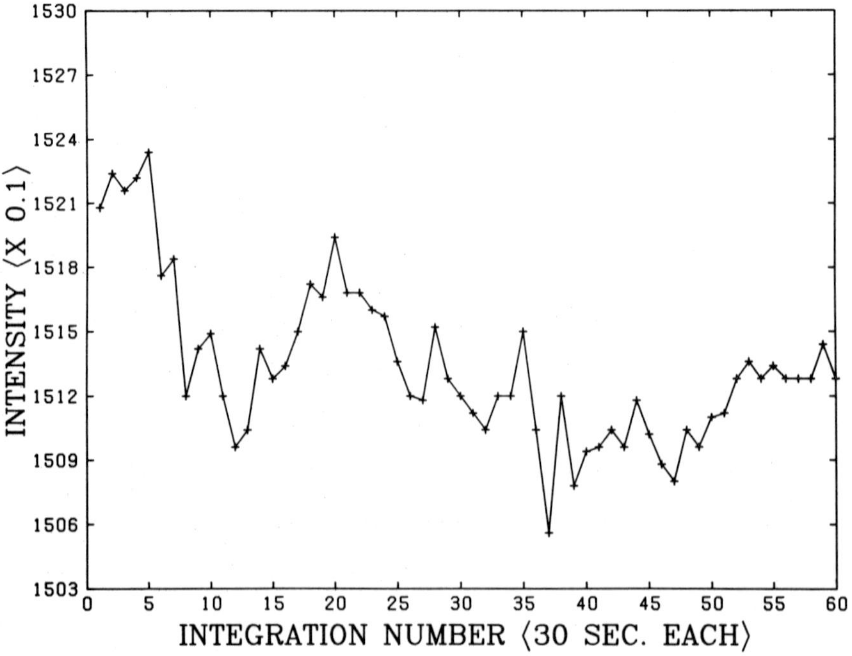

FIGURE II. RUN SEQUENCE PLOT OF INTEGRATED ARGON EMISSION. 394.9nm; 0.25% RSD.

The third and major source of variability is the solution introduction system, namely, the nebulizer and its associated spray chamber. Clearly, the nebulizer is the main contributor to the system behavior shown in Figure 1.

In our laboratory, the data are obtained with the instrument minicomputer but are filed and analyzed by a large central computer. In detecting the presence of random and nonrandom errors we use statistical routines in a data analysis package called DATAPAC [2]. A four-

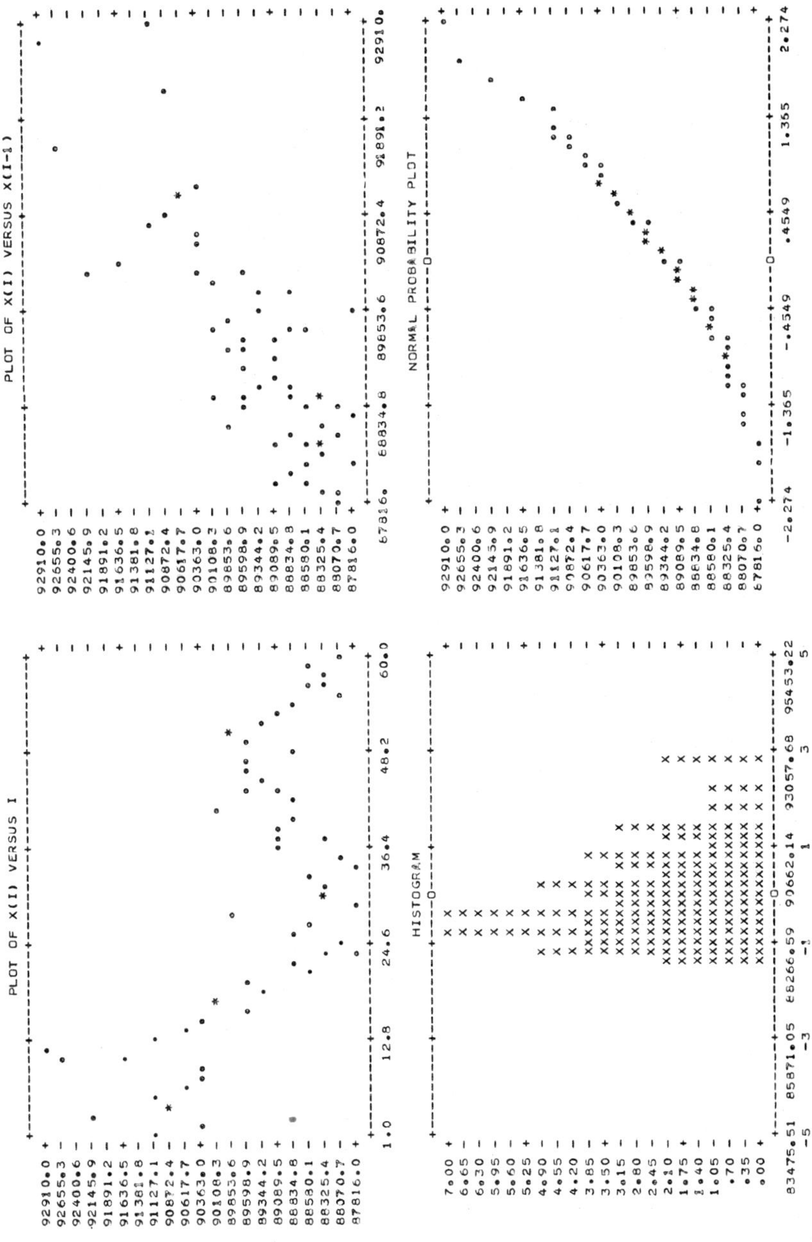

FIGURE III. FOUR-PLOT STATISTICAL ANALYSIS OF DATA IN FIGURE I.

plot routine [3] which is available in our central computer OMNITAB [4] library is shown in Figure 3. Analysis of the run sequence in Figure 1 is presented in four quadrants of a line printer plot. The upper left plot is a repeat of the run sequence itself and a histogram of the intensity data is plotted in the lower left quadrant. The upper right plot is a lag-one correlation plot; that is, one intensity data point is plotted against the previous intensity data point. Any discernable patterns in this plot indicate nonrandom error caused by sequential

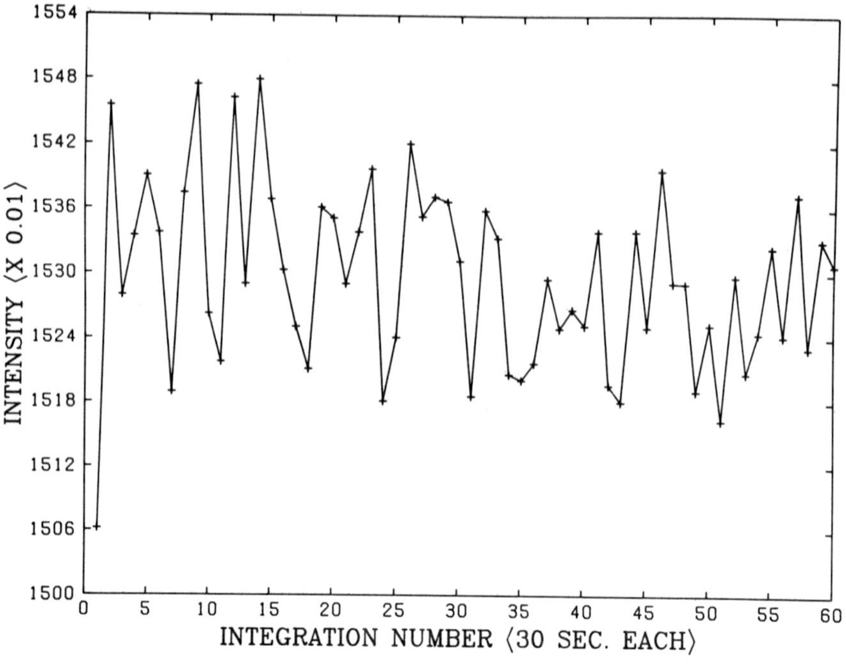

FIGURE IV. RUN-SEQUENCE PLOT OF INTEGRATED COPPER EMISSION. 327.4nm; aspiration; 0.56% RSD.

correlation of the data. The lower right plot is a test for normal probability distribution of the data. From the analysis given in Figure 3 it is evident that the impression of nonrandom variability in the data is a valid one.

Another example of a run sequence comparable to that of Figure 1 is given in Figure 4. Occasionally, aspiration with the nebulizer produces data which do not display any apparent drift. This is verified by the statistical plots in Figure 5, although the occurrence of a single outlying data point is evident. It is not known at this time what conditions are necessary to consistently produce randomly variable data.

One additional problem with direct aspiration of a sample by the nebulizer is the effect of head pressure changes. If the change in solution level during aspiration is significant compared to the total height of the solution above the end of the aspiration tube, negative intensity drift as shown in Figure 6 will result. For the case shown, an intensity decrease of approximately 10 percent occurred between integration numbers 48 and 54. During this period of time the solution head changed from 2.0 cm to 1.2 cm. Various constant-head devices can be employed although it is more convenient in our laboratory to use solutions of sufficient volume to minimize this effect. The use of a peristaltic pump with the pneumatic nebulizer has not proven to be a satisfactory solution to the head pressure effect. The pulsations inherent in this type of pump result in periodic intensity fluctuations as shown in Figure 7. The variance of the data increases up to 20 percent RSD as the integration period is shortened to values approach-

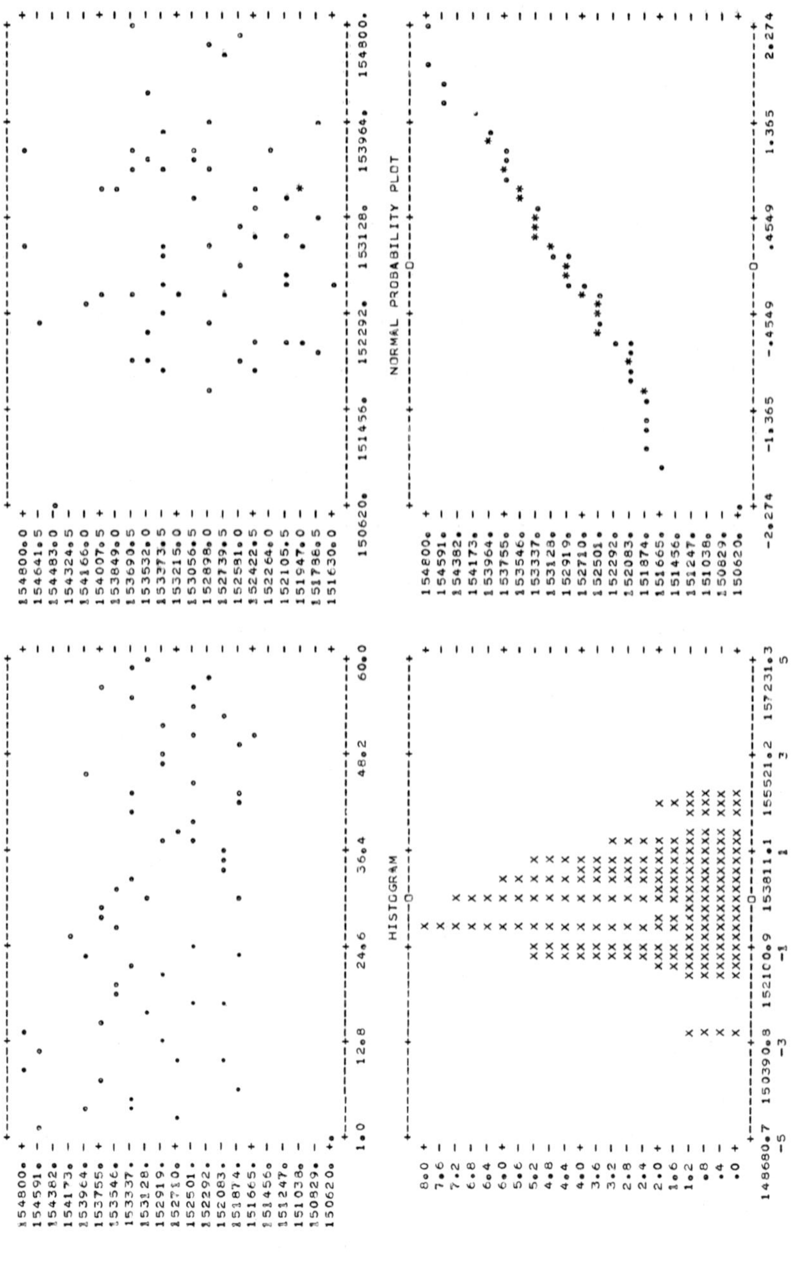

FIGURE V. FOUR-PLOT STATISTICAL ANALYSIS OF DATA IN FIGURE IV

ing a pulse period (about 1 sec). The use of a peristaltic pump with an ultrasonic nebulizer has not been tested in our laboratory.

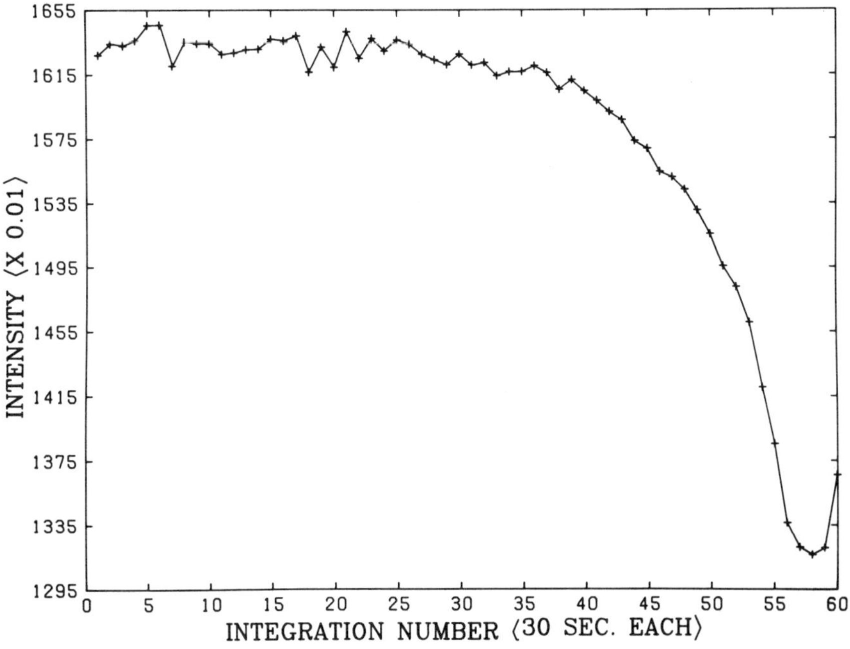

FIGURE VI. INSTRUMENTAL DRIFT DUE TO DECREASING SAMPLE HEAD PRESSURE. Copper, 327.4nm.

Instrumental drift may or may not introduce systematic error in an analysis. If the integration time for a given intensity measurement is long compared to the periodicity of the drift then the measurement is subject to random error. If this is not the case, several integrations may be taken over a long enough time period to ensure that this average will produce random and normally distributed results. However, the data plotted in Figure 1 indicate that the periodicity of the drift is sufficiently long that other techniques are needed to randomize results of replicate analyses.

One approach to reducing the effects of instrumental drift is to ratio an element intensity to that of an internal standard. It has been observed that the typical direct intensity variance of 1 to 2 percent RSD (cf. Fig. 1) observed for a number of elements can be reduced to

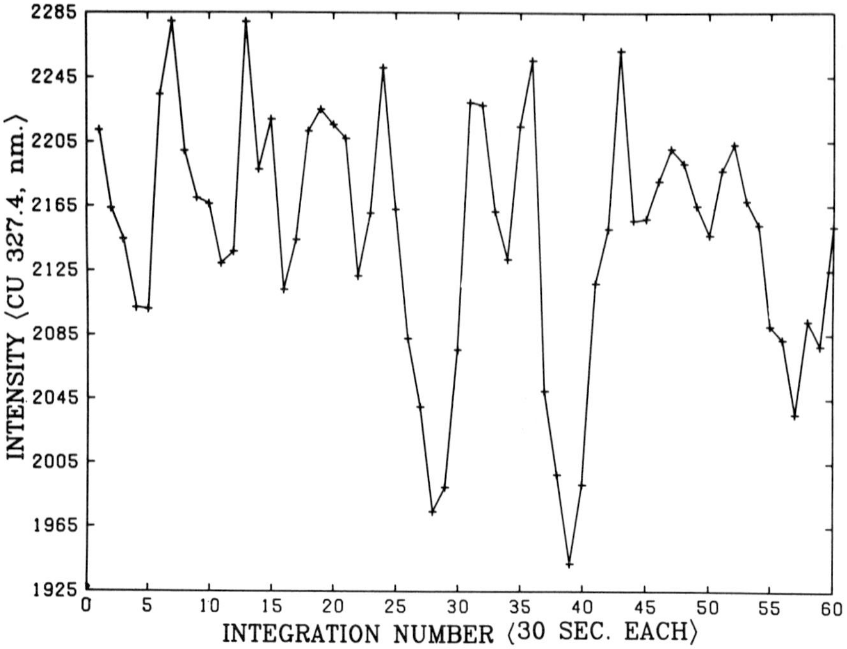

FIGURE VII. INTENSITY PULSATIONS FROM A PERISTALTIC PUMP USED WITH A PNEUMATIC NEBULIZER. 3.6% RSD.

less than 0.5 percent RSD by concurrent ratioing of the intensities to the proper internal standard element. Selection of the proper internal standard element and spectral line is an ongoing concern in analytical atomic spectroscopy. The various criteria that have evolved from excitation energy, ionization energy, and partition function considerations [5] can be advantageously applied to selection of sample-standard pairs of lines. However, since one of the most attractive features of the ICP technique is its applicability to simultaneous multielement analysis, selection of the best standard element and spectral line for all sample elements of interest is not always possible or practical. This is especially true when polychromators with fixed channels are used. When intensities of a particular element in a run sequence are ratioed to an internal standard channel which is not well matched, the total variance may be greater than that associated with the direct intensity data.

Another approach which is often employed to reduce the nonrandom error due to drift is to restandardize the instrument by aspirating standards before and after the sample. To evaluate the effectiveness of this procedure a sample and standard which were aliquots of the same solution were examined in order to eliminate chemical and spectral interferences. The aliquots were aspirated successively allowing four intensity integrations (two minutes) for washout and four integrations for an average intensity for each aliquot. The data are plotted as a run sequence in Figure 8. The four data points which were averaged for each aliquot are connected by a solid line. If the first aliquot is

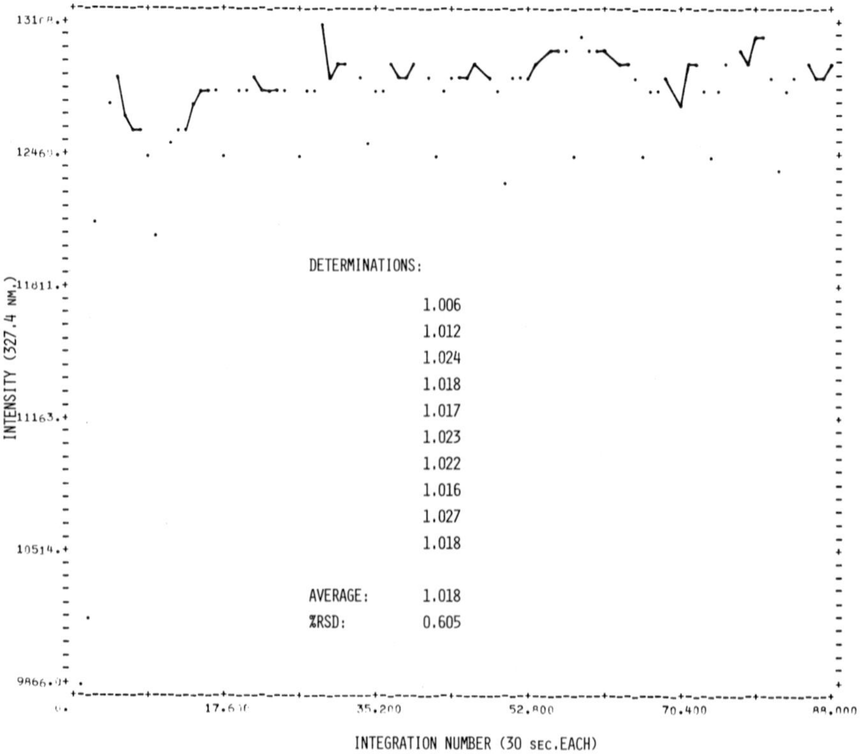

FIGURE VIII. INITIAL STANDARD-SUBSEQUENT SAMPLE RUN SEQUENCE.

taken as the standard containing 1.000 μg/mL copper and each successive aliquot is taken as a sample, the resulting determinations as tabulated in the figure can be calculated. Since the sample and standard are identical in this example, the blank is also identical for each. Therefore raw intensities can be used to establish the slope of the calibration curve with the y-intercept assumed to be zero. Obviously for real samples background or blank corrections are needed to obtain net intensities and more than a single standard is needed to obtain the calibration curve. In this case instrumental drift causes a 1.8 percent error in the average of ten replicate determinations.

If the instrument is restandardized before and after each sample the situation is improved. The same data are plotted in Figure 9, except that every other aliquot is taken as the standard. Each determination is calculated from the average of the previous and succeeding

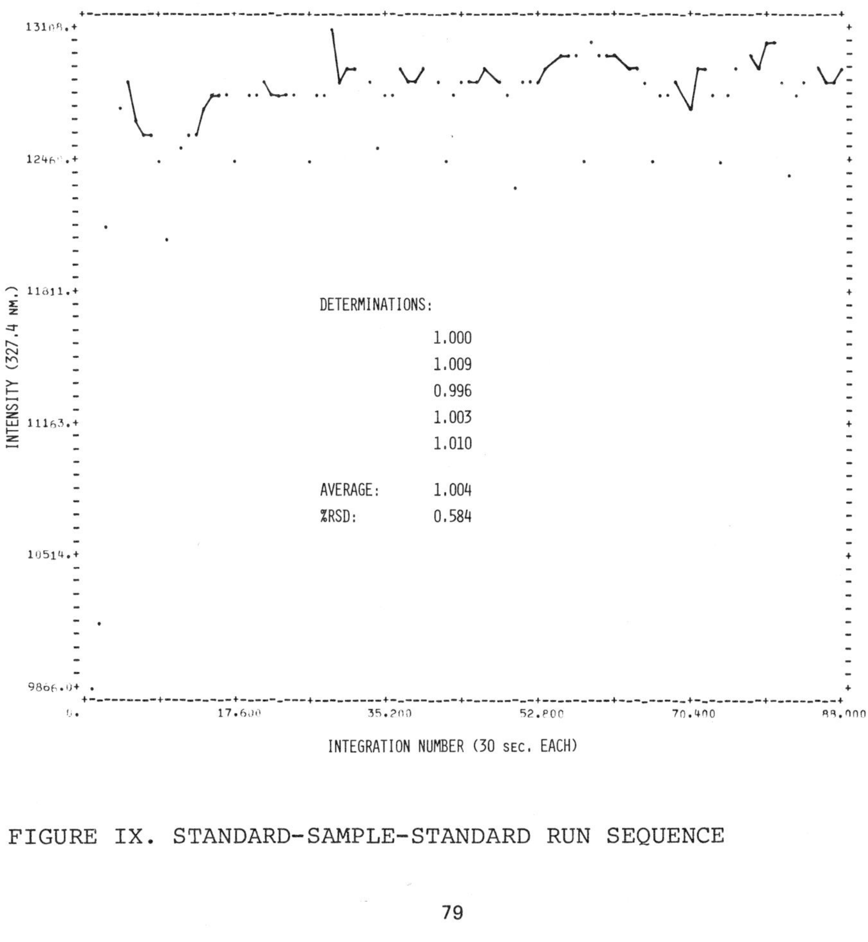

FIGURE IX. STANDARD-SAMPLE-STANDARD RUN SEQUENCE

standards. The determinations are tabulated as before and a reduction of error in the average to 0.4 percent is indicated in the figure. Here, only five determinations are made in the same time period as the previous case.

CONCLUSION

Instrumental drift in the ICP technique may introduce systematic error in an analysis. Total instrumental variability without solution introduction is better than 0.5 percent RSD over a time period of one-half hour. The nebulizer and associated spray chamber contribute up to three times this level of variability when added to the system. A one page four-plot routine was used as a convenient first-pass analysis of the data to separate random and nonrandom errors. When time-dependent nonrandom errors are present, ratioing intensities or bracketing samples and standards may reduce their effect on the final result.

For practical analyses the number of samples may preclude the time-consuming effort required in matrix matching the standards. In such cases some technique of background correction must be utilized which will most likely lengthen the total time of an analysis. These considerations must balance those of instrumental drift to arrive at a run sequence which will be cost effective at the level of accuracy desired.

REFERENCES

[1] Boumans, P. W. J. M. and de Boer, F. J., Spectrochim. Acta, Part B, 27, 291 (1972).

[2] Filliben, James J., DATAPAC: A Data Analysis Package. Proceedings of the Ninth Interface on Computer Science and Statistics, Prindle, Weber and Schmidt, Inc., Boston, Mass., 212 (1977).

[3] Filliben, "Testing Basic Assumptions in the Measurement Process", in Validation of the Measurement Process, James R. DeVoe, Editor, ACS Symposium Series 63, Washington, DC, p. 30 (1976).

[4] Hogben, David, Peavy, Sally T. and Varner, Ruth N., NBS Magnetic Tape 1: Omnitab II Magnetic Tape and Documentations Parcel, NBS Publication Announcement TRA 71-125, Dept. of Commerce, January 1972.

[5] Barnett, W. B., Fassel, V. A. and Kniseley, R. N., Spectrochim. Acta, Part B, 23, 643 (1968).

ANALYSIS OF FOODS AND RELATED MATERIALS USING INDUCTIVELY COUPLED PLASMA SPECTROSCOPY

John W. Jones and Kenneth W. Boyer

INTRODUCTION

Recognition of the significant roles of metals at trace levels in biological systems has led to development of countless analytical methods designed to facilitate analyses of various types of biological and environmental materials for low levels of these elements. Accurate quantitation of the metal content of foods and food-related materials is of particular importance because of the relatively large proportion of metals ingested by the population via food products. Continuous monitoring of foods for a number of important nutritional and toxic elements is necessary to ensure that trace metal nutritional balance is maintained, and that food contamination from any source is minimized. Atomic absorption spectroscopy (AAS), currently the most commonly used single-element determinative technique for this monitoring, provides the bulk of recent data on trace metals in foods. Recently, however, multielement techniques such as neutron activation analysis

TABLE IA. TYPICAL RANGES OF SELECTED ELEMENTS IN FOODS[a] (µg/g)

	LOW	HIGH
>100 µg/g		
Ca	20	20,000
K	20	50,000
Mg	10	1,000
Na	10	70,000
P	10	20,000
1 µg/g - 100 µg/g		
Cu	0.01	10
Fe	0.1	50
Mn	0.01	15
As	0.001	100
Se	0.001	2
Zn	0.05	100
0.1 µg/g - 1 µg/g		
Cd	N.D.[b]	0.5
Cr	N.D.	0.5
Pb	N.D.	1
Ni	N.D.	1
<0.1 µg/g		
Be	N.D.	0.002
Co	N.D.	0.10
Sb	N.D.	0.05
Te	N.D.	0.002
V	N.D.	0.05

a. References (1,9,17,18).
b. Not detectable by most state-of-the-art instrumentation: generally <0.001 µg/g.

(NAA), x-ray fluorescence (XRF), and inductively coupled plasma emission spectroscopy (ICP) have been successfully applied to trace metal analyses of food-related materials (1-3). The ICP currently offers the the greatest analytical potential for these materials in terms of cost, sensitivity, and simplicity.

Morrison recently outlined the potential of the ICP to quantitate a number of important metals in biological materials (4). We have chosen a similar approach to illustrate some anticipated performance characteristics of the plasma discharge in analysis of food and related materials. Table 1A lists a number of metals (and metalloids) that are of interest to the U.S. Food and Drug Administration (FDA) from a nutritional and/or toxicological standpoint. The elements are categorized as to their typical ranges (μg/g) in foods. The overall range is from less than 1 ng/g, e.g., Be, up to 2-3% for Ca or K. It should be pointed out that Table 1A includes data on most food types, and thus the fairly wide ranges reflect high concentrations of certain elements in particular foods, e.g., As in seafoods. Table 1B lists ICP quantitation limits in foods for several elements from the previous table. Sample quantitation limits are defined as 50 times the solution detection limits (d.l.), i.e., 1 g of sample in 10 ml of analysis solution times 5 d.l. Finally, Table 1C shows the elements having overlapping ranges and ICP quantitation limits in food matrices. Consequently, only about 2/3 of the elements of interest are regularly quantifiable in foods via direct ICP measurement. The remaining elements generally require a separation and concentration step prior to ICP measurement; these elements will not be considered in detail here. The majority of the elements in Table 1C may be readily determined by direct ICP measurement provided that appropriate sample preparation procedures, instrumental parameters, and

TABLE IB. ESTIMATED ICP DETECTION LIMITS AND SAMPLE QUANTITATION LIMITS FOR SELECTED ELEMENTS IN FOODS

ELEMENT	ICP DETECTION LIMIT, μg/ml	QUANTITATION[a] LIMITS IN FOODS, μg/g
As	0.1	5.0
Be	0.001	0.05
Ca	0.0005	0.025
Cd	0.002	0.1
Co	0.005	0.25
Cr	0.005	0.25
Cu	0.009	0.45
Fe	0.009	0.45
Mg	0.006	0.3
Mn	0.001	0.05
Ni	0.01	0.50
P	0.2	10.0
Pb	0.05	2.5
Se	0.04	2.0
Sb	0.1	5.0
Te	0.1	5.0
V	0.01	0.5
Zn	0.005	0.25

a. Calculated quantitation limits are based on 1 g of food material, ashed and diluted to 10 ml; actual values change with dilution factor.

spectral considerations are evaluated and optimized prior to the measurement step. The following sections outline some of the important factors influencing determination of several elements in foods and related materials via ICP measurement.

TABLE IC. ELEMENTS REGULARLY QUANTIFIABLE IN FOODS

ELEMENT	ANALYTICAL WAVELENGTH, nm
As[a]	193.7
Ca[b]	393.3 and 370.6
Cd[a]	226.5
Cu	324.7
Fe	259.9
K[c]	----
Mg	279.5
Mn	257.6
Ni[a]	341.4
P	214.9
Pb[a]	220.3
Se[a]	196.0
Zn	206.1

a. These elements are often below ICP quantitation limits, but are sometimes present at measurable levels in certain food groups, e.g., Cd in leaf vegetables.

b. Two Ca lines are used to adequately cover the wide range of this element in food matrices.

c. FDA ICP spectrometer not equipped with K analytical channel.

APPARATUS

ICP measurements were performed on a Jarrell-Ash Model 975 Plasma-AtomComp equipped with a 0.75 meter focal-length direct-reading spectrometer, a 25 μm fixed entrance slit (3mm aperture), a wavelength modulation background corrector Model 90-555 "Spectrum Shifter," and appropriate photomultiplier tubes.

Spectral scans were obtained via a Jarrell-Ash 0.5 meter Ebert monochromator equipped with variable entrance and exit slits.

A cross-flow pneumatic nebulizer (5) operated at 1 liter argon per minute at 20 psi was used for sample introduction into the ICP. The argon plasma, operated at 15 l argon/min, was maintained by a Plasma-Therm Model HFS-2000D RF generator at 1100 watts forward power at a fixed frequency of 27.12 MHz.

SAMPLE MINERALIZATION PROCEDURES

Sample mineralization procedures which are compatible with direct ICP determinations of the elements listed in Table 1C must meet several requirements: The ashing procedure must (a) be adaptable to the ICP sample introduction device, commonly a pneumatic nebulizer; (b) be sufficiently fast to accommodate the high sample throughput requirement of most laboratories; and (c) be capable of quantitatively recovering all of the analytes from the food matrix.

The first of these criteria listed above is particularly important.

Currently used pneumatic nebulizers are prone to instability or clogging when solutions containing fairly low percentages of dissolved solids are aspirated. Although the exact salt tolerance varies with the chemical nature of the dissolved material and with nebulizer design (indeed, among different units of similar design) approximately 1% (w/v) dissolved solids is normally an upper limit for commonly used pneumatic ICP nebulizers. For many pneumatic nebulizers 1% may be optimistic. Consequently, the first consideration in the analytical sequence for foods must be proper selection of sample size and dilution volume. Most foods contain between 0.5% and 6% inorganic ash by weight. Therefore, it is convenient to choose a minimum dilution of 1 g of food material per 10 ml of ash solution. Some foods may require a greater or lesser dilution. The 1 to 10 dilution factor was used in calculations of the quantitation limits in Table 1B. Further discussion of the effects of dissolved solids relative to nebulizer performance is presented in the next section of this report.

Either dry ash or wet ash sample mineralization procedures may be used to satisfy the sample dissolution requirements. The usual advantages and disadvantages of each must be weighed in order to arrive at the optimum procedure. We have used both techniques in our work. Both techniques can provide suitable sample solutions for ICP analysis. Table 2 shows comparative ICP analyses of a green bean homogenate which was ashed by two different procedures. Ten 10 g replicates of the homogenate were wet ashed using a mixture of redistilled nitric and perchloric acids (4:1), then diluted to 50 ml. In addition, 10 similar

TABLE II. WET ASH-DRY ASH COMPARISON FOR GREEN BEANS

ELEMENT	WET ASH,[a] $HNO_3/HClO_4$	DRY ASH[a] H_2SO_4 ASH AID
	µg/g	
Ca	279 ± 4	263 ± 5
Cu	0.26 ± 0.01	0.24 ± 0.02
Fe	4.03 ± 0.09	3.68 ± 0.07
Mg	126 ± 2	121 ± 2
Mn	1.57 ± 0.02	1.49 ± 0.03
P	123 ± 1	117 ± 6
Zn	1.54 ± 0.02	1.49 ± 0.03

a. Values shown are mean ± 1 standard deviation for ten 10 g samples.

samples from the same homogenate were dry ashed at 475°C using a sulfuric acid ashing aid added to minimize volatilization losses of several elements, dissolved in 10% (v/v) perchloric acid (to maintain a similar matrix to the wet ashed samples), and diluted to 50 ml with 10% (v/v) perchloric acid. All samples were then analyzed by ICP. Background correction using wavelength modulation was employed for Cu, Fe, Mn and Zn. With the exception of Fe, essentially the same values were obtained for the quantitated elements using the two ashing procedures. Other food products behaved similarly. The overall speed of each technique is quite comparable if large numbers of samples must be ashed. There are, however, potential problems associated with each ashing technique relative to recovery of many of the analytes.

WET ASHING WITH PERCHLORIC ACID. For the most part, nitric acid/perchloric acid oxidation of food materials is an effective sample mineralization technique, resulting in quantitative recovery of most elements in Table 1C. Use of perchloric acid, however, can lead to two serious problems in the presence of moderate potassium concentrations (i.e., > 2000 μg/g). The first is the coprecipitation of many analyte elements by potassium perchlorate. The $KClO_4$ precipitate occurs primarily with samples having a fairly high potassium content, e.g., 1-3% as in dried plant materials such as soy flour or dried leaf vegetables. Table 3 illustrates the potential error. Ten 10 g samples of NBS Orchard Leaves (SRM-1571) were wet ashed with a nitric/perchloric acid mixture. The NBS certified potassium value for the Orchard Leaves is 1.47 $\pm$ 0.03%. A substantial precipitate remained after digestion, severely affecting the Fe, Mg, Mn and P values. The NBS values are reported here to one standard deviation for comparison purposes. Results from a similar set of Orchard Leaves dry ashed with H_2SO_4 ashing aid gave considerably better agreement with the NBS values. The problem of potassium perchlorate coprecipitation of analyte metals has been pointed out previously (6).

The second problem associated with use of perchloric acid involves samples containing between 2,000 and 10,000 μg K/g. In such cases, the dissolved potassium perchlorate in the sample digest solution is near saturation. This leads to erratic nebulizer behavior during aspiration

TABLE III. EFFECT OF $KClO_4$ PRECIPITATE FORMATION DURING $HNO_3/HClO_4$ ACID DIGESTION OF 1 g NBS ORCHARD LEAVES

ELEMENT	THIS WORK[a,b]	NBS [a]
	μg/g	
As	9 ± 2	11 ± 1
Ca	1.96% ± 0.11%	2.09% ± 0.15%
Cd	N. Q.	0.11 ± 0.01
Cu	11 ± 1	12 ± 0.5
Fe	233 ± 23	(270)
Mg	0.51% ± 0.02%	0.62% ± 0.01%
Mn	80 ± 5	91 ± 2
Ni	N. Q.	1.3 ± 0.1
P	0.14% ± 0.01%	0.21% ± 0.005%
Pb	40 ± 6	(44)
Zn	23 ± 2	25 ± 1.5

a. Values expressed as mean ± 1 standard deviation.

b. Ten samples.

of the solution into the ICP and can eventually lead to total clogging of the nebulizer due to formation of a deposit inside the glass capillary of the crossflow nebulizer. Dilution of the solution reduces the effect but results in poorer quantitation limits. The problems associated with potassium perchlorate can be minimized by carrying out the wet digestion using a mixture of nitric, perchloric and sulfuric acids rather than nitric and perchloric acids alone. The perchloric acid is distilled off, reducing the potential for chemical or physical interferences from insoluble perchlorates.

DRY ASHING. Dry ash sample mineralization has been successfully used as a solubilization step prior to ICP determination of a number of elements in biological tissues, plant materials, and other complex matrices (6,7). Most often the ashing is carried out at about 500°C. Many elements of direct nutritional or toxicological importance are quantitatively recovered with this procedure. Notable exceptions are As, Se, Pb, Cd and perhaps Cu and Fe (8). Various ashing aids have been used to minimize volatilization and vessel retention losses of these elements. Sulfate, for instance, as sulfuric acid or ammonium sulfate is sometimes used to convert volatile Pb and Cd salts to thermally stable forms (9). Magnesium oxide and magnesium nitrate ashing aids can be used to retain As during dry ashing of foods at 500°C (10). However, use of ashing aids can cause problems either in the ashing step or during the ICP determination. For example, sulfuric acid, while providing the means for good dry ashing recoveries for Pb and Cd, etc., can lead to formation of an acid-insoluble residue during dry ashing, causing substantial losses of Fe, Mg, Mn, Ni, P, Zn and probably others, if present. Table 4 shows ICP data obtained on five replicate samples of a canned vegetable soup homogenate; sulfuric acid was used as an ashing aid to quantitatively recover Pb and Cd for determination by electrochemical analysis at levels not detectable by ICP. Samples A, B and C contained no residues;

TABLE IV. EFFECT OF RESIDUE FORMATION DURING DRY ASHING WITH H_2SO_4 ASHING AID

ELEMENT	µg/g SAMPLE (WEIGHT IN g)				
	A (11.65)	B (13.68)	C (13.97)	D (14.98)	E (17.31)
Ca	119	119	118	118	117
Cu	0.15	0.20	0.21	0.18	0.19
Fe	20.0	19.9	19.8	14.0	15.7
Mg	84.1	83.4	82.7	69.4	70.8
Mn	1.61	1.64	1.62	1.49	1.49
Ni	1.03	1.03	1.06	0.92	0.88
P	259	260	257	211	218
Zn	8.54	8.55	8.52	8.01	8.00

D and E were slightly larger samples and contained insoluble white residues. Samples of the same homogenate (up to 30 g) ashed without sulfuric acid contained no residue. Surprisingly, Ca, whose sulfate salt is relatively insoluble, did not appear to be affected by the residue. Subsequent x-ray fluorescence analysis of residues from these and similar samples showed insufficient sulfur to suggest a sulfate salt, but did show relatively high phosphorus content. The sulfuric acid apparently leads to formation of insoluble pyrophosphates of potassium, magnesium and other elements, a phenomenon mentioned by Gorsuch (8). Whatever the chemical nature of the dry ash residue, its presence should not be attributed solely to insoluble siliceous materials.

Use of alkali and alkaline earth salts or oxides as ashing aids usually provides clean ashes, but can cause nebulization problems if the final ash solution contains more than about 1% (w/v) dissolved solids.

Additional data for wet ashing and dry ashing of foods and biological materials prior to ICP determination have been reported (11-13). While there are certainly advantages to both techniques, the diversity of food matrices dictates that no single sample preparation procedure can apply equally well to all samples for subsequent ICP measurement of several elements.

BACKGROUND AND MATRIX CONSIDERATION

Criteria for analytical line selection for ICP detection of trace elements in foods have been reported (14). As with most atomic spectroscopic techniques, the most desirable line exhibits a low detection limit and minimal interference from matrix elements. An ICP source usually offers a number of analytically useful emission lines (either atomic or ionic) for a particular element, and as a result, a sensitive line, relatively free from major interferences, is often available. The lines listed in Table 1C were chosen primarily because of their fairly high sensitivity, and their minimal interferences from the major matrix elements in foods: Ca, Na, K, P, and Mg.

The term "interference-free" has sometimes been used to describe the characteristics of the ICP. Certainly, the plasma discharge provides an attractive environment for high intensity emission with minimum chances for chemical interferences; for measurements of trace element concentrations near the detection limits, however, even minor contributions to the analytical signal from spectral line or background interferences can result in serious errors if no correction procedures are employed. Larson et al (15) have pointed out that apparent background shifts caused by matrix elements may, in fact, be due to stray light within the spectrometer, rather than to actual changes in the background near the analytical line. Whatever the source of the detected radiation, an appropriate correction for background, such as that which can be obtained by wavelength modulation, should be employed. If apparatus for wavelength modulation background correction is unavailable, matrix matching of calibration standards to the sample solution can partially compensate for the major matrix interferences. Indeed, for some food-related materials it is desirable to use both background correction and matrix matching. Table 5 illustrates the accuracy attainable in a sample type representing perhaps a worst-case example of a food matrix. The data in Table 5 are comparative analyses for lead (Pb) in several acid-digested bone meal food supplements. Analyses were performed by ICP and anodic stripping voltammetry (ASV). The ICP determinations were performed using matrix-matched standards consisting of 10,000 μg Ca/ml, 5,000 μg P/ml, and 200 μg Mg/ml in 10% (v/v) $HClO_4$. Wavelength modulation background correction was made using the Jarrell-Ash Model 90-555 "Spectrum Shifter." Agreement between the two

TABLE V. ICP AND ASV DETERMINATION OF Pb IN ACID-DIGESTED BONE MEAL

	μg Pb/ml [a]	
SAMPLE NO.	ICP[b]	ASV[c]
1	0.16	0.17
2	0.16	0.16
3	0.25	0.24
4	0.22	0.21
5	0.10	0.08
6	0.42	0.44

a. Values given are in μg Pb/ml for approximately 2.5 g of original dry material digested in nitric/perchloric acid and brought to volume of 50 ml. ICP and ASV measurements were made on the same solutions.

b. ICP values obtained on wavelength-modulated background-corrected system against matrix matched standards: 10,000 μg Ca/ml; 5,000 μg P/ml; and 200 μg Mg/ml.

c. ASV measurements were made at pH ~5 on 2.0 ml original solution diluted to 12 ml with deionized water.

measurement techniques is remarkably good considering that the ASV measurement has a detection limit about three orders of magnitude lower than does the ICP measurement. Figure 1 illustrates the magnitude of the background shift under the 220.3 nm Pb line relative to the background with deionized water. Measurement of the bone meal samples against aqueous standards without background correction on the direct reading spectrometer resulted in as much as 375% positive error.

It should be pointed out that the matrix matching used for the bone meal samples served not only to help compensate for background shifts

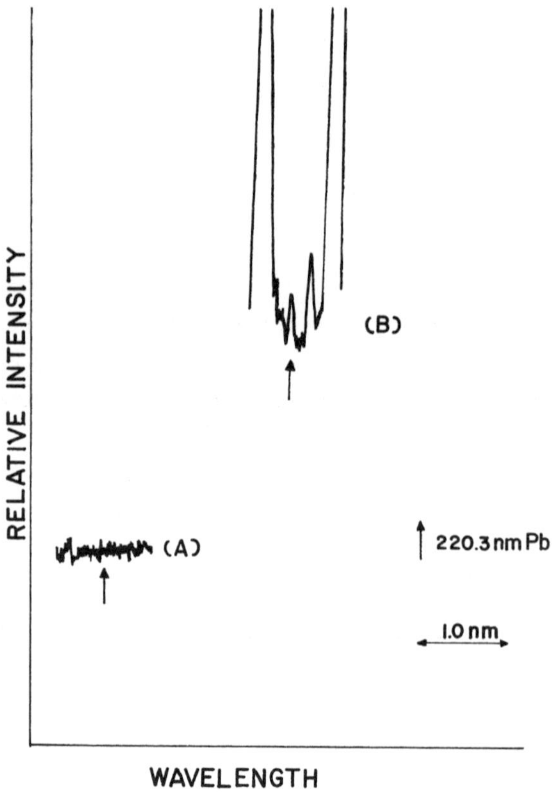

FIGURE I. WAVELENGTH SCANS IN THE VICINITY ON THE Pb II (220.3 nm) LINE FOR (A) DEIONIZED WATER AND (B) ACID-DIGESTED BONE MEAL. Note the large increase in background intensity under the Pb line and the strong Ca line on either side of the Pb. Background correction was made at ~ 0.3Å on the high wavelength (right) side of the Pb line.

and possible ionization interferences, but to more or less equalize solution uptake rates on the cross-flow pneumatic nebulizer for samples and standards (i.e., viscosity). Solution uptake effects due to dissolved salts are not generally severe in most foods as long as the sample preparation procedures outlined earlier are closely followed. However, when high concentrations of dissolved salts are encountered, matrix matching can help compensate for the effects.

An obvious disadvantage of matrix matching is that in most food-related analyses, the matrix element (Ca, Na, K, P, and Mg) concentrations are unknown. In such cases, it is necessary to first determine the concentrations of the matrix elements in the samples prior to matrix matching and measurement of the trace elements. For high solution concentrations of any of these major elements (e.g., > 500 μg/ml) in the original sample solution, it is advisable to perform one or two dilutions in an appropriate acid solvent with subsequent ICP measurement to ensure that no significant viscosity problems exist between the original sample solutions and the calibration standards. If differences in analyte response are found, dilutions of the original samples are required to minimize the effects prior to ICP measurement.

Obviously, the matrix influence on solution uptake rate is greatly dependent upon the type of nebulizer used. The effects reported here were observed to occur with a cross-flow pneumatic nebulizer. Similar effects may or may not occur with different nebulizer designs. Suddendorf and Boyer have recently developed a nebulizer which significantly improves precision with high salt content solutions by eliminating the problem of clogging (16). However, this nebulizer is not yet in general use.

PRECISION

Overall analytical precision for any analysis is ultimately affected by each manipulative and measurement step performed by the analyst. The sample preparation procedure, which includes all operations such as homogenization, weighing, etc., and the instrumental procedure contribute to the final precision. The relative standard deviation for repetitive ICP measurement of a single sample solution for most elements at the concentrations generally present in foods is around 1.0%. Moreover, sample preparation usually contributes no more than an additional 1-5% RSD to the overall precision. As with most techniques, precision deteriorates as the levels of the analytes approach the quantitation limits, i.e., five times the solution detection limit. Typical precision for the analysis of three foods for seven elements is presented in Table 6. Several 10 g samples of each thoroughly homogenized commodity were wet ashed with a mixture of nitric and perchloric acids and diluted to 50 ml before ICP measurement. With the exception of Fe in the peach sample, all RSD values were less than 5%. These three sample types also illustrate the variability of the concentration of a particular element among various sample types, e. g., Ca, P, and Zn.

Samples that prove particularly difficult to ash either by wet or

TABLE VI. TYPICAL OVERALL ANALYSIS PRECISION FOR SOME REPRESENTATIVE FOODS[a,b]

	MEAN ± 1 STANDARD DEVIATION, μg/g		
ELEMENT	CANNED INFANT FORMULA, N=9	GLASS PACKED GREEN BEANS, N=10	CANNED PEACHES, N=3
Ca	820±8(0.98)	279±4(1.43)	27.4±0.2(0.73)
Cu	1.07±0.02(1.87)	0.26±0.01(3.85)	0.26±0.01(3.85)
Fe	21.6±0.1(0.46)	4.03±0.09(2.23)	1.9±0.3(15.8)
Mg	97.6±0.8(0.82)	126±2(1.59)	54.2±0.1(0.18)
Mn	0.09±0.00(0)	1.57±0.02(1.27)	0.27±0.01(3.70)
P	560±4(0.71)	123±1(0.81)	76±2(2.63)
Zn	8.57±0.08(0.93)	1.54±0.02(1.30)	0.38±0.01(2.63)

a. $HNO_3/HClO_4$ digestion.

b. Values in parentheses are RSD(%).

dry mineralization procedures typically yield poorer overall precision due to volatilization losses or to incomplete oxidation of the organic matrix. Such imprecision is, of course, independent of the ICP measurement. Similarly, samples containing nonhomogenizable particulate material may lead to poor precision if precautions are not taken to ensure homogeneity. These observations point out that the sample preparation procedures, and not the ICP measurement, contribute the major portion of the total analytical imprecision for the determination of most quantifiable elements in foods.

APPLICATIONS OF THE ICP TO OTHER FOOD-RELATED MATERIALS

The ICP spectrometric analysis system currently provides an excellent tool for direct analysis of foods for about 8-10 elements. But in addition to foods, the ICP can be used for elemental analysis of many "food-related" substances. Broadly defined, "food-related" refers to: (1) any material which may enter into foods, either directly from approved food additives, or indirectly from food packaging, storage, or serving materials; (2) food supplements such as vitamins and minerals; and (3) fertilizers, animal feeds, and pesticides which may ultimately introduce nutritional or toxic elements into foods.

The analytical complexity for substances such as those listed above is often compounded by the fact that many elements not normally encountered at high concentrations in foods may predominate in food-related products. For example, Table 7 shows concentrations of several elements in two very different food-related substances, a mineral supplement and a sewage sludge sample (sludges are sometimes used as fertilizers). The concentrations of most elements shown in Table 7 are 2-3 orders of magnitude higher than levels found in foods. Judicious selection of sample sizes (to ensure both homogeneity and ICP quantitation ability for trace elements) is required for the ICP determination. Similarly, reevaluation of spectral interferences, both line and background, must be performed to minimize the ICP measurement error. Indeed, for very high concentrations of Fe, a preset direct reading-analytical line for Co (238.9 nm) may be rendered useless, or at best questionable for trace measurements even with computer-assisted interference corrections. Thus, while the ICP analysis system can be adequately employed for a variety of food-related sample types, careful recalibration and sample characterization procedures should be maintained.

TABLE VII. PARTIAL ELEMENTAL CONTENT OF A MINERAL SUPPLEMENT AND A SEWAGE SLUDGE SHOWING HIGH CONCENTRATIONS OF POTENTIALLY INTERFERING METALS

ELEMENT	MINERAL SUPPLEMENT[a]	SEWAGE SLUDGE[b]
	μg/g	
Al	4,600	16,100
Ca	15,000	49,000
Cu	1,350	430
Fe	10,400	43,800
Mg	42,500	35,300
Mn	304	1,040
P	65	13,400
Ti	490	1,570
Zn	592	1,050

a. ICP determination on nitric/perchloric acid digest.

b. Relatively high concentrations of Cd, Co, Cr, and Pb were also present in the sludge but are not shown here.

SUMMARY

Use of an ICP discharge for emission spectroscopic measurements can provide an excellent means of simultaneous quantitation of 8-10 elements at the levels at which they are normally present in food materials. The number of elements that can be quantitated depends on the dilution factor used, which in turn is a function of the ash content of the food material; 1 g of material per 10 ml of analysis solution is a convenient dilution factor. Interferences arising from matrix elements are generally quite subtle, but can be significant at or near the quantitation limits. Sample preparation can introduce significantly greater error and imprecision to the overall analysis than does the ICP measurement step alone. Though wet ashing and dry ashing can both be used for sample mineralization for the ICP, each technique has potential problems that must be minimized before the ICP can yield accurate results. The ability to rapidly and simultaneously quantitate a number of elements of interest in foods and related materials makes the ICP an attractive alternative compared to other techniques currently in use.

REFERENCES

1. Tanner, J.T. and Friedman, M.H., J. Radioanal. Chem., 37 (1977) 529.

2. Irons, R.D., Schenk, E.A., and Giangue, R.D., Clinical Chemistry, 22 (1976) 2018.

3. Sobel, H.R., Kniseley, R.N., Sutherland, W.L. and Fassel, V.A., Paper No. 339, presented at the 26th Pittsburgh Conference on Analytical Chemistry and Applied Spectrometry, Cleveland, Ohio, March 1975.

4. Morrison, G.H., ICP Information Newsletter, 3, No.5 (1977) 165.

5, Kniseley, R.N., Amenson, H., Butler, C.C., and Fassel, V.A., Applied Spectrometry, 28 (1974) 285.

6. Dahlquist, R.L. Knoll, J.W., and Irons, R.D. Paper No. 80, presented at the 3rd Annual Federation of Analytical Chemistry and Spectroscopy Societies, Philadelphia, Nov. 1976.

7. Scott, R.H. and Strasheim, A., Analytica Chimica Acta, 76 (1975) 71.

8. Gorsuch, T.T., The Destruction of Organic Matter, Pergamon Press, New York (1970).

9. Jones, J.W., Gajan, R.J., Boyer, K.W., and Fiorino, J.A., J. Assoc. Offic. Anal. Chem., 60 (1977) 826.

10. Hundley, H.K. and Underwood, J.C., J. Assoc. Offic. Anal. Chem., 53 (1970) 1176.

11. Ward, A.F. Paper No. 99, presented at the 89th Annual Meeting of the Association of Official Analytical Chemists, Washington, D.C., Oct. 1975.

12. Nixon, D.E. and Fassel, V.A., Paper No. 224, presented at the 27th Pittsburgh Conference on Analytical Chemistry and Applied Spectroscopy, Cleveland, Ohio, March 1976.

13. Gunn, A.M., Kirkbright, G.F., and Opheim, L.M., Anal. Chem., 49 (1977) 1492.

14. Winge, R.K. and Fassel, V.A., Paper No. 24, presented at 4th Annual Federation of Analytical Chemistry and Spectroscopy Societies, Detroit, Nov. 1977.

15. Larson, G.F., Fassel, V.A., Winge, R.K., and Kniseley, R.N., Appl. Spectros., 30 (1976) 384.

16. Suddendorf, R.F. and Boyer, K.W., Paper No. 395, presented at the 29th Pittsburgh Conference on Analytical Chemistry and Applied Spectroscopy,Cleveland, Ohio, March 1978.

17. Watt, B.K. and Merrill, A.L., "Composition of Foods", U.S. Dept. of Agriculture Handbook, No. 8, Wash., D.C. (1963).

18. Fiorino, J.A., Jones, J.W., and Capar, S.G., Anal. Chem. 48 (1976) 120.

RECENT GEOCHEMICAL APPLICATIONS OF THE INDUCTIVELY COUPLED PLASMA

Danold Golightly

INTRODUCTION

The range of applications of the inductively coupled plasma (ICP) to chemical analysis problems in the earth sciences has expanded considerably in the past 3 - 4 years. Rocks and minerals (1-8), concentrates (9-10), sulfide ores (11), coal and coal ash (12), soils (13-16), water (17, 18), and plants (19, 20) have been analyzed recently by ICP spectrometry. Much on-going work with the ICP in laboratories involved in exploration geology or concerned with basic geochemical research is expected to enrich the literature during the next several years.

Geological materials are both a diverse and a complex sample group. Thus, approaches to sample treatment preceding analysis must be keyed to the specific matrix and to the objectives of the analysis. Sample treatment includes all grinding, sieving, fusion, and dissolution procedures necessary to prepare a sample for analysis.

TABLE I. CONTAMINANTS FROM GRINDING (21, 22)

Agate	Alumina	Alumina-Ceramic	Boron Carbide	Tungsten Carbide
Si	Al	Al	B	W
	Cr	Cu		Co
	Fe	Fe		Ti
	Ga	Ga		
	Zr	Li		
		Ti		
		B		
		Ba		
		Co		
		Mn		
		Zn		
		Zr		

SAMPLE TREATMENT AND CONTAMINATIONS

A. Grinding and Sieving

Every bulk rock sample must be reduced to a powder by grinding procedures prior to chemical analysis. All grinding procedures for hard rocks add contaminants, and thus, subsequent chemical decomposition methods must be effective for relatively coarse powders, that is, powders produced by minimal grinding. Most rock samples can be pulverized to 100 mesh (approximately 150 μm) without unacceptable contamination occurring. Each of the currently available grinder-substrate materials, e.g., agate, alumina, boron carbide, tungsten carbide, etc., has its own set of contaminating elements that can be introduced into a sample. These contaminants and common grinding materials are surveyed in Table I.

TABLE II. CONTAMINANTS FROM SIEVING (21, 22)

Nylon	Brass	Stainless Steel
-	Cu	Co
	Pb	Cr
	Sn	Fe
	Zn	Mn
		Ni

Nylon sieves are preferred over brass or stainless steel because of the appreciable contamination by Co, Cu, Fe, Mn, Ni, Pb, Sn, and Zn from the latter two materials (Table II).

B. Fusion and Dissolution

The decomposition of most silicate rocks is based on the formation of volatile silicon tetrafluoride with hydrofluoric acid or on the formation of alkali metal silicates and aluminates by reaction with molten alkali metal salts. Carbonate rocks and others can be decomposed by common acids, such as hydrochloric, perchloric, and sulfuric acids. Teflon acid-bombs are being used more frequently in dissolution procedures for minerals and rocks. Currently, the lithium metaborate - tetraborate flux is in widespread use for decomposition of silicate rocks. The resulting fusion bead is readily soluble in dilute nitric acid. Obviously, other fluxes, such as sodium peroxide, alkali metal hydroxides and fluorides, potassium pyrosulfate, and sodium carbonate - borate mixtures, can be used. Regardless of the choice of decomposition procedure, the possibility of contamination from reagents and the potential for nebulizer effects and interelement effects arising from the high salt concentrations associated with fusion methods must be considered. Salt concentrations up to 0.4 % are tolerated by some common pneumatic nebulizer systems (7). Higher concentrations and higher injection rates can be accommodated by ultrasonic nebulizers, but frequently, undesirable desolvation interferences and salt deposition within the plasma torch occur (23). These limitations are imposed by nebulizer - aerosol transport systems and do not reflect a tolerance threshold for solids by the ICP.

APPLICATIONS

In principle, multielement analysis of any geochemical sample is possible if the appropriate chemistry for dissolution, preconcentration, etc., can be contrived. A survey of some recent applications of the ICP is presented in Table III. The sample type, elements determined, approximate concentration ranges for analysis, and dissolution methods given in this table provide a concise overview of current ICP methodology. The original references provide both the details necessary for the successful analysis of the individual sample matrices and the interpretation of data essential to a judgement of the merits of each analytical approach.

A. Exploration Geology

In general, approaches to exploration geology include multielement chemical analysis methods that are applied to systematic searches for mineral deposits. Anomalous concentrations of the elements associated with mineralization are sought in rock, soil, sediment, vegetation, groundwater, and the atmosphere. Barringer (24) recently introduced an airborne exploration technique that utilizes the ICP for analysis of vegetation and soil particles collected on an adhesive tape from an air sampler flown over regions being characterized. More conventional samples have been analyzed by Scott and colleagues in investigations of trace elements in soil (16) and orchard leaves (19) and of uranium in rocks (5, 6). Air-dried samples (16) were treated with a 9:1 mixture of perchloric and nitric acids, and the supernatant liquid analyzed for Cu, Zn, Ni, Co, and Pb. Good agreement was observed between results obtained for Cu and Zn by ICP spectrometry and by atomic absorption spectrometry. Typical coefficients of variation ranged from 5-10% for element concentrations above 20 μg/ml, where measurements were made on a 0.5 m Ebert spectrometer with a 1 s time constant for the measurement circuit. Six varieties of orchard leaves were analyzed with the same spectrometer - ICP system for Fe, Mn, Cu, Al, B, and Zn (19). Again, the results compared favorably with those from other methods, including atomic absorption, x-ray fluorescence, and d.c. arc emission spectrometry. In the case of uranium in core samples and ores, Scott et al (5, 6) found that an ion line, U (II) 3782.84 Å, provided optimal emission from an 800 watt argon

TABLE III. APPLICATIONS OF THE INDUCTIVELY COUPLED PLASMA TO ANALYSIS OF GEOCHEMICAL SAMPLES

Sample	Elements determined	Concentration range (μg/ml)	Dissolution method	Reference
Rocks (granite, granodiorite, andesite, basalt, diorite, peridotite)	Na, K, Mg, Ca, Fe, Al, Ti, P, Be, Sr, Ba, Zr, V, Cr, Mn, Co, Ni, Cu, Zn.	---	Acid digestion with mixture of $HF-HClO_4-HNO_3$.	3, 4
Rocks (granites, basalt, diorite, serpentine)	P	0.6 - 46	$HF + HClO_4$ mixture.	8
Rocks (ores)	U	0.2 - 170	Diluted $HNO_3 + H_2SO_4$ mixture.	6
Exploration Samples (drill cores)	U	0.2 - 200	Diluted $HNO_3 + H_2SO_4$ mixture.	5
Sulfide Concentrates (copper, lead, zinc)	Ag, As, Au, Ca, Cd, Co, Mg, Mn, P, Sb, Ti, etc.	---	Na_2O_2 fusion, melt leached with dilute HNO_3	9, 10

TABLE III. (Continued)

Sample	Elements determined	Concentration range (μg/ml)	Dissolution method	Reference
Sulfide Ores (heavy mineral separates from drill cores)	Ag, Al, As, Ba, Bi, Ca, Cd, Co, Cu, Fe, K, Mg, etc.	50 ppm - 0.1% (generalized range)	Aqua Regia, taken to dryness and redissolved in 10% HCl.	11
Coal and Coal Ash	Fe, Ti, Ca, Al, Mg, Na, Cr, Se, Pb, etc.	0.01 - 600 (generalized range)	HNO_3 - $HClO_4$, HF - $HClO_4$, and HF - HNO_3 mixtures.	12
Soils	Cu	0.1 - 40	Acid extraction with mixture of $HClO_4$ - HNO_3.	16
	Zn	20 - 255		
	Ni	10 - 240		
	Co	5 - 35		
	Pb	10 - 500		
Water	Ag, Bi, Cd, Cu, Nb, Ni, Pb, Sn, W.	---	None	17, 18
Orchard Leaves (apple, apricot, grape, pear, peach, plum)	Fe	3 - 7	Ash at 500° C; dissolve ash in dilute HNO_3.	19
	Mn	1 - 7		
	Cu	0.1 - 0.8		
	Al	3 - 10		
	B	0.5 - 2		
	Zn	0.5 - 1		

plasma. This line was free from spectral interferences by naturally concomitant elements, and enabled detection of uranium at a concentration of 0.1 μg/ml. The method, which entails a 2-hour dissolution procedure in a 1:1 mixture of nitric and sulfuric acids, enables convenient determinations of uranium in the concentration range from 0.002 - 2 % U_3O_8 in rock and ore samples. In a batch mode, this approach should be competitive with the x-ray fluorescence procedure of Clark and Pyke (25).

Burman (8) and Kirkbright et al (13) have determined phosphorus in rocks and soils, respectively. The P (I) 2136.18 Å line gives a detection limit for phosphorus of 0.08 μg/ml and enables the determination of phosphorus in rocks from 0.006 to 1 % (0.6 - 46 μg/ml in solution) and in soils from 10 to 80 ppm (0.25 - 2 μg/ml in solution). In addition, Kirkbright et al (13) have shown that the S (I) 1820.37 Å line enables the determination of sulfur at concentrations from 2 to 1000 μg/ml in aqueous solutions. The potential for the determination of total sulfur in important geochemical matrices, e.g., coal, is indicated by these data.

B. Basic Geochemical Studies

An example of more basic geochemical studies is found in an application of the ICP to determining zirconium - hafnium ratios in zircons and rocks. These elements have long been recognized as a classic example of a geochemically coherent pair whose close association is explained by their chemical similarity and their nearly identical ionic radii (26). Precise Zr/Hf concentration ratios assist the geologist in

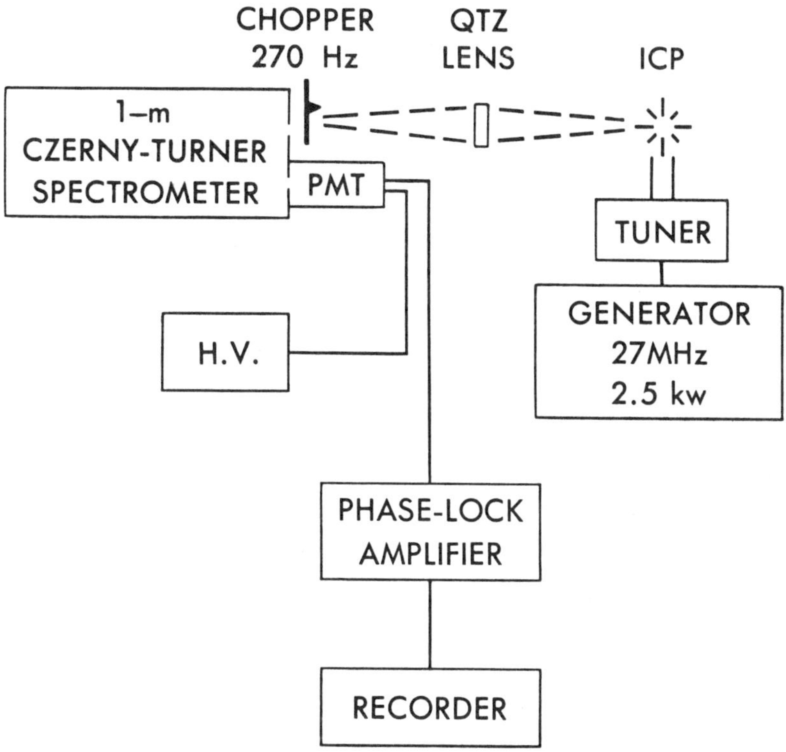

FIGURE I. DIAGRAM OF INSTRUMENTATION FOR SPECTROMETRY OF THE ICP

the interpretation of magmatic differentiation processes. These ratios are being determined for zircon solutions in 10 % sulfuric acid (27). The Zr (II) 3399.35 Å - Hf (II) 3399.80 Å line-pair intensity ratio is measured by spectral scanning with the spectrometer system diagrammed in Figure 1.

ROLE OF THE ICP IN GEOCHEMICAL ANALYSIS

In the near future, the ICP will be used increasingly to meet the requirements of exploration geochemistry (28), especially for minerals related to energy needs and for minerals required by the metallurgical industry. The technique can be tailored to economically analyze materials on a multielement basis (29-32) and because of its inherent precision, complements x-ray fluorescence spectrometry in determinations of elements in the parts-per-million concentration range (31, 32). The good determination limits for many elements that are not readily excited in more ordinary emission spectrometric sources suggests possibilities for basic geochemical studies previously unattainable because of the lack of adequate analytical capabilities. The ICP is expected to assume a greater role in all aspects of geochemical analysis.

REFERENCES

(1) E. Dekalb, E. M. Layton, and V. A. Fassel, "Inductively Coupled Plasma Optical Emission Spectrometry: Application to the Determination of the Major and Minor Elemental Composition of Minerals," 17th Annual Rocky Mountain Spectroscopy Conference, Denver, Aug. 5, 1975.

(2) R. Crawford and A. Ward, "Analytical Performance of the ICAP as Applied to Mineralogical Matrices," 22nd Spectroscopy Symposium, Montreal, Oct. 26-29, 1975.

(3) S. E. Church, "Independent Analysis of the U. S. Geological Survey Standard Rocks Using an Inductively Coupled Plasma Quantometer," Pittsburgh Conference on Analytical Chemistry and Applied Spectroscopy, Cleveland, Feb. 28 - Mar. 4, 1977.

(4) S. E. Church, "Analysis of the U. S. Geological Survey Standard Rocks Using an Inductively Coupled Plasma Quantometer," 19th Annual Conference on Analytical Chemistry, Denver, Aug. 1-3, 1977.

(5) R. H. Scott and M. L. Kokot, ICP Inform. Newsletter, 1, 34(1975).

(6) R. H. Scott, A. Strasheim, and M. L. Kokot, Anal. Chim. Acta, 82, 67(1976).

(7) F. N. Abercrombie and M. D. Silvester, "Applications of the Argon RF Plasma to Real Sample Systems," Pittsburgh Conference on Analytical Chemistry and Applied Spectroscopy, Cleveland, Mar. 3-5, 1975.

(8) J. O. Burman, ICP Inform. Newsletter, 3, 33(1977).

(9) A. E. Watson, G. M. Russell, and G. Balaes, Report No. 1815, South African National Institute for Metallurgy, Apr. 15, 1976.

(10) A. E. Watson, G. M. Russell, and G. Balaes, "The Commissioning of an Induction-Coupled Plasma System and Its Application to the Analysis of Copper, Lead, and Zinc Concentrates," IUPAC Int. Symp. Anal. Chem. Expl. Mining Proc. Matls., Johannesburg, South Africa, Aug. 23-27, 1976.

(11) S. E. Church, "The Analysis of a Sulfide Ore Suite Using the ICPQ," Applied Research Laboratory, Sunland Laboratory Applications Report, Aug., 1977.

(12) A. F. Ward, "Analysis of Coal and Coal Ash Samples Using the Inductively Coupled Argon Plasma," Jarrell-Ash Applications Report No. 7, 1976.

(13) G. F. Kirkbright, A. F. Ward, and T. S. West, Anal. Chim. Acta, 62, 241(1972).

(14) A. Danielson, "Application of ICP Spectrometry to Samples of Geological and Environmental Interest," Pittsburgh Conference on Analytical Chemistry and Applied Spectroscopy, Cleveland, Mar. 1-5, 1976.

(15) R. L. Dahlquist and J. W. Knoll, "Soil and Biological Tissue: ICP Multielement Determinations," Pittsburgh Conference on Analytical Chemistry and Applied Spectroscopy, Cleveland, Feb. 28 - Mar. 4, 1977.

(16) R. H. Scott and M. L. Kokot, Anal. Chim. Acta, 75, 257(1975).

(17) C. T. Apel, T. M. Bieniewski, L. E. Cox, and D. W. Steinhaus, Report No. LA-6751-MS, Los Alamos Scientific Laboratory of the University of California, Los Alamos, New Mexico, Mar. 1977.

(18) C. T. Apel, T. M. Bieniewski, L. E. Cox, and D. W. Steinhaus, ICP Inform. Newsletter, 3, 1(1977).

(19) R. H. Scott and A. Strasheim, Anal. Chim. Acta, 76, 71(1975).

(20) F. Brech, ICP Inform. Newsletter, 1, 195(1976).

(21) G. Thompson and D. C. Bankston, Appl. Spectrosc., 24, 210(1970).

(22) H. Bennett, Analyst (London), 102, 153(1977).

(23) P. W. J. M. Boumans, ICP Inform. Newsletter, 3, 71(1977).

(24) A. R. Barringer, "Multi-element Trace Analysis Applied to Airborne Geochemical Exploration," IUPAC Int. Symp. Anal. Chem. Expln. Mining Proc. Matls., Johannesburg, South Africa, Aug. 23-27, 1976.

(25) N. H. Clark and J. G. Pyke, Anal. Chim. Acta, 58, 234(1972).

(26) H. Klemic, D. Gottfried, M. Cooper, and S. P. Marsh, "Zirconium and Hafnium," chapter in D. A. Brobst and W. P. Pratt, Eds., "United States Mineral Resources," U. S. Geol. Survey Prof. Paper 820, 713-722(1973).

(27) D. W. Golightly and F. O. Simon, U. S. Geol. Survey J. Res. (1977). (Manuscript in preparation.)

(28) C. H. James, "Analytical Requirements in Exploration Geochemistry," IUPAC Int. Symp. Anal. Chem. Expln. Mining Proc. Matls., Johannesburg,

South Africa, Aug. 23-27, 1976.

(29) P. W. J. M. Boumans, Philips Tech. Rev., 34, 305(1974).

(30) P. W. J. M. Boumans, Proc. Analyt. Div. Chem. Soc., 14, 143(1977).

(31) G. Tolg, Naturwissenschaften, 63, 99(1976).

(32) I. Rubeska, Analusis, 4, 314(1976).

A NEW MULTIELEMENT TECHNIQUE FOR THE COLLECTION AND ANALYSIS OF AIRBORNE PARTICULATES IN AIR QUALITY SURVEYS

F. N. Abercrombie, M. D. Silvester,
A. D. Murray, and A. R. Barringer

ABSTRACT

AIRTRACE[R] is an integrated air quality surveillance system consisting of a particulate sizing module, sample impingement collection deck, pulsed laser sampling playback deck, induction coupled plasma emission spectrometer, and computer/software/hardware package. The design and sensitivity of the system is such that only 2 to 15 seconds of collection are required for most unpolluted areas, allowing for the inclusion of background areas, as well as polluted areas, in an air quality survey. A sampling rate of 400 sample pairs per hour permits the study of rapid atmospheric quality changes in stationary surveys or fine three dimensional geographic resolutions in mobile surveys.

The potential utility of AIRTRACE has been demonstrated during a survey of urban particulate sources. In this survey, a 600 square mile area at three altitudes was sampled for two particulate size fractions over a five hour period. A total of 2,000 samples were collected, providing a recovered flight path distance of 200 miles. Element abundances were reported in map form, using graphic computer processing.

INTRODUCTION

For the last decade Barringer Research has taken an active role in the development and manufacture of sophisticated analytical instrumentation for the analysis of trace constituents in the atmosphere. This instrumentation has often been remote sensing to obtain the advantages of airborne surveys as well as to provide a means of by-passing property trespass restraints for air quality studies.

Previous instrumentation has included COSPEC (SO_2 and NO_2) and GASPEC. Several years ago, a corporate priority was assigned to the development of remote sensing instrumentation capable of providing multielement analysis of airborne particulates. The usefulness of such a system for air quality measurements is immediately obvious. However, in a geophysical exploration mode, real benefits are also derived from the multi-element analysis of airborne particulates. Through several complex physical processes, particulate plumes are found in the vicinity of many types of mineral deposits; thus, a multi-element analysis of the particulate plume provides qualitative elemental characterization data for subsurface mineral deposits as complimentary support data for classical electromagnetic geophysical anomalies located within the survey area.

Several requirements were defined after reviewing the problem of multi-element analysis of airborne particulates and these are itemized in Table 1. These requirements can be explained as follows. If collection is conveniently accomplished, the maximum number of samples can be obtained within a given time period. A short sampling interval will permit rapid time-varying atmospheres to be monitored without losing the time structure through time integration which is commonly encountered with high-vol filters; a similar advantage will be obtained in mobile sampling situations where a maximum of detail of the particulate composition is desired along with an interest to maintain normal flying or driving speeds. The capability to make particle size cuts

TABLE I. REQUIREMENTS FOR COLLECTION AND ANALYSIS OF AIRBORNE PARTICULATES

: Short Sampling Interval

: Fractionate Particle Sizes

: Sensitive to Background Levels

: Rapid Analysis

: Multi-Element

: Matrix Independent

: Portable

: Automated

permits the analysis of multi-element distributions between particle sizes and resultant mean airborne lifetime. Sensitivity to background levels will yield maximum information. A rapid analysis will provide feedback information during the course of the sample collection period. Multi-element capabilities will also maximize the information obtained as well as permit near absolute analysis. Matrix independence is required to provide meaningful results for the varied types of samples encountered. The collection system must be lightweight and self-sustained if it is to be truly portable, and finally, to achieve many of the preceding requirements, the system must be automated.

COLLECTION INSTRUMENTATION

At the beginning of the development, a new collection system was considered because none of the existing systems available satisfied all the collection requirements in Table 1. A schematic diagram of the collection system is shown in Figure 1. The primary sample collectors are "air scoops" which capture and transfer the air samples into either louvered inertial collectors (aerodynes) or cyclones. Both collectors are inertial devices. Lightweight cyclones can be manufactured to handle several hundred cubic feet of air per minute with good collection efficiency for the requisite particle sizes and the louvered collectors can readily handle air volumes of as much as 12,000 $feet^3$ /minute in an airborne system. Either separation provides an air sample stream which is enriched with airborne particulates. The enriched air stream is

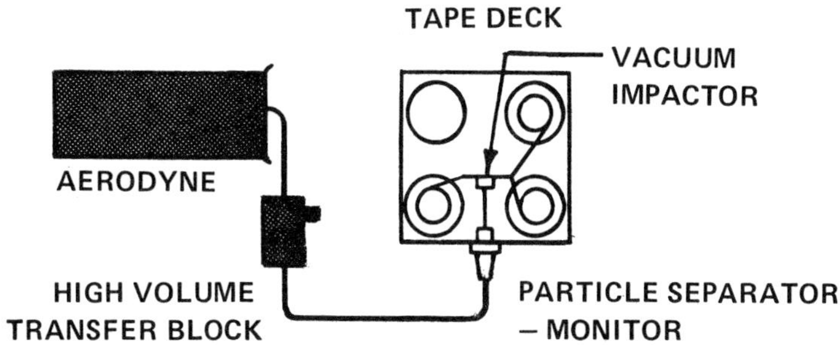

FIGURE I. SCHEMATIC DIAGRAM OF AIRTRACE[R] COLLECTION SYSTEM

passed through two transfer blocks which further enrich the particulate to air concentration ratio by inertial processes. The airborne particulates are then vacuum impacted onto the adhesive surface of a flexible Mylar type tape. Particulates approaching 1μ are collected with appropriate design. (The collection efficiencies rapidly decrease for particulates less than 1μ.) The tape is stationary during the collection interval and is rapidly advanced to a fresh collection site at the end of each collection interval of from 5 to 60 seconds. As the tape is advanced, it is automatically marked with a sample locating mark and a teflon cover tape is laid over the sample to prevent contamination and facilitate the subsequent handling of the sample tapes. This procedure produces a series of circular particulate spots (about 3 mm in diameter) on the tape adhesive, with the coarse and fine particles adjacent to one another. The analytical cleanliness of the fresh sample tape and cover tape are protected with a clean air atmosphere obtained from a pressurized nitrogen cylinder. Static charges are dissipated by a polonium strip. The collection deck, cyclones, aerodynes, collection monitoring instruments, ancillary controls and power pack permit stationary, van-mounted or airborne installations; the latter can be in either fixed-wing or rotary aircraft.

ANALYTICAL INSTRUMENTATION

The development of a tape deck system for the collection of airborne particulates was relatively straightforward. However, the development of a "playback" deck was complicated by the task of removing 10 to 500μg of sample from an adhesive tape surface and transporting the resulting aerosol to a multi-element instrument without contamination or matrix interferences. A schematic diagram of the tape "playback" is given in Figure 2. A pulsed 17 megawatt CO_2-TEA laser was chosen to effectively release the airborne particulates from the sample tape. The ejected particulates (Figure 3) are entrained in the flowing argon stream passing through a specially designed glass cell (Figure 4) and swept into a radio frequency induction coupled argon plasma (ICAP). Approximately 80-85% of the ejected sample is carried to the ICAP. The design of the cell prevents cross contamination between adjacent sample pairs as well as the ability to open and close the playback assembly without extinquishing the ICAP. When the playback deck is opened the argon escapes out across the tape, but the argon is forced into the cell and the ICAP when the playback deck is sealed.

In the automated mode, the sample tape first advances the sample into the target zone (focused image) of the laser. The laser fires.

After 10 seconds, a front-surfaced mirror in the laser path rotates slightly to reflect the laser beam image onto the second sample spot and the laser fires. After 10 seconds, the mirror rotates to its original position and the sample tape advances to a fresh pair of spots beginning the cycle again. The direct-reading plasma spectrometer integration commences coincident with the laser pulse and continues for

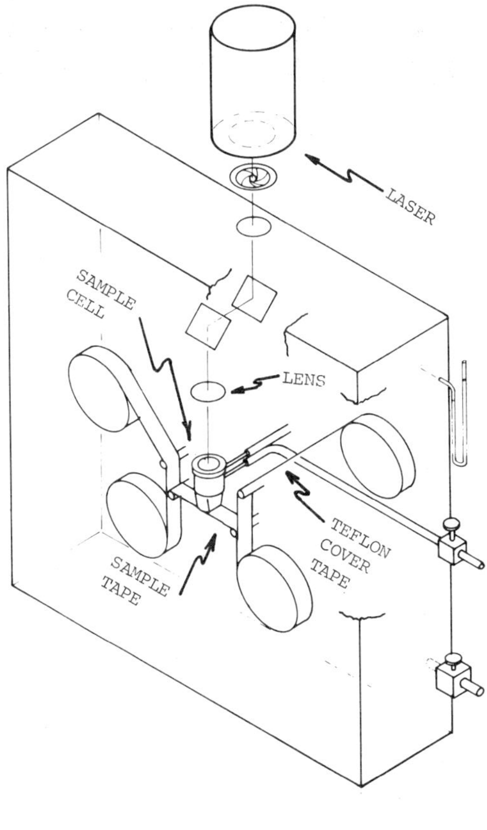

FIGURE II. SCHEMATIC DIAGRAM OF SAMPLE ANALYSIS TAPE DECK AND LASER OPTICS

FIGURE III. PHOTOGRAPH OF THE LASER-SAMPLE PLASMA AND INCANDESCENT PARTICULATE EJECTA

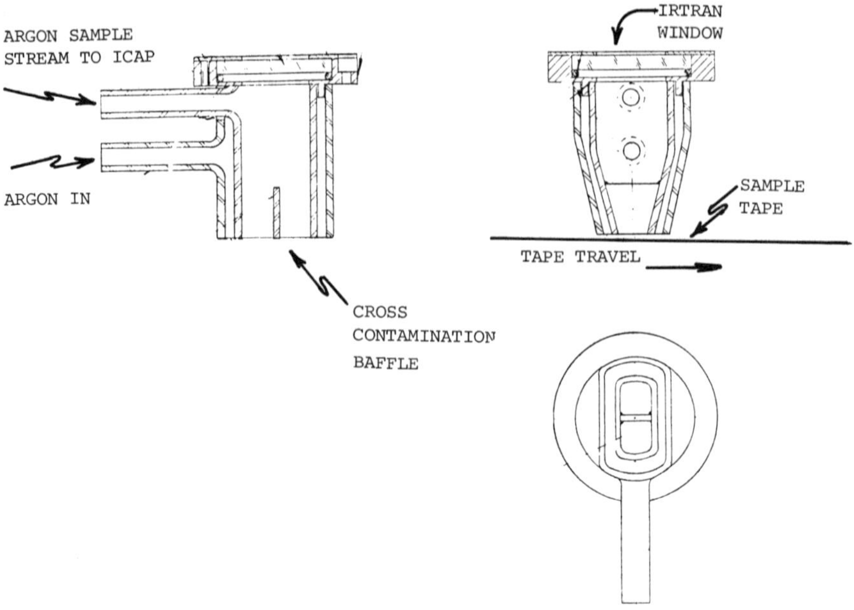

FIGURE IV. SAMPLE CELL FOR COLLECTION OF LASER EJECTA

3 seconds. During that time interval approximately 80% of the transported material is swept through the plasma. The analytical rate is limited to a 10 second interval by the operating speed of the calculator.

EMISSION SPECTROMETER

The induction coupled argon plasma system used in this work was an Applied Research Laboratory QA137 direct-reading spectrometer with a 25 element slit array. The output from the spectrometer digital voltmeter was interfaced through a Hewlett-Packard 9821 programmable calculator which provided magnetic tape cassette storage and thermal printer hard copy record. The ICAP was operated at 1600 watts forward power at an observation height of 16 mm above the load coil. Coolant, plasma and aerosol gas flowrates were 10, 1 and .9 litres per minute respectively.

ANALYTICAL RESULTS AND DISCUSSION

Initial studies included an evaluation of the collection efficiencies of the tape deck for various particle sizes and also the linearity of the analytical system. The induction coupled argon plasma had demonstrated several orders of magnitude linear response in many laboratories, but this particular application with a laser had not been studied. USGS soils were carefully sprinkled on the adhesive tape to provide different loadings of the soils upon tape and were analysed by the instrumentation. The millivolt values of element plotted vs.

element (Figures 5 and 6) demonstrate the linearity of response for iron vs. silicon, and manganese vs. silicon obtained for USGS GXR 6. These results are typical of results obtained for all element pairs for several USGS standards.

The LASERTRACE instrumentation is calibrated for geological materials. A voltage ratio, VS_j, is defined for each element to the voltage for silicon for standards and samples:

$$VS_j = \frac{(mV)_j}{(mV)Si} \qquad (1)$$

(The spectrometer signal response is a voltage reading).

If the voltage ratio is obtained from a standard, it is also possible to define a standard concentration ratio RS_j of the concentration of element j to the concentration of the element silicon in the standard.

$$RS_j = \frac{ppm_j}{ppm_{Si}} \qquad (2)$$

These two ratios RS_j and VS_j can be used to define a proportionality factor (K_j) for the jth element channel relative to silicon.

$$K_j = RS_j/VS_j \qquad (3)$$

A voltage ratio VX_j from a sample is also defined as the ratio of the voltage of the element j to the silicon voltage.

$$VX_j = \frac{(mV)_j}{(mV)_{Si}} \qquad (4)$$

The proportionality constant K_j times the sample voltage ratio VX_j gives a concentration ratio RX_j for the ratio of the element j concentration to the silicon concentration in the sample.

$$RX_j = VX_j \cdot K_j \qquad (5)$$

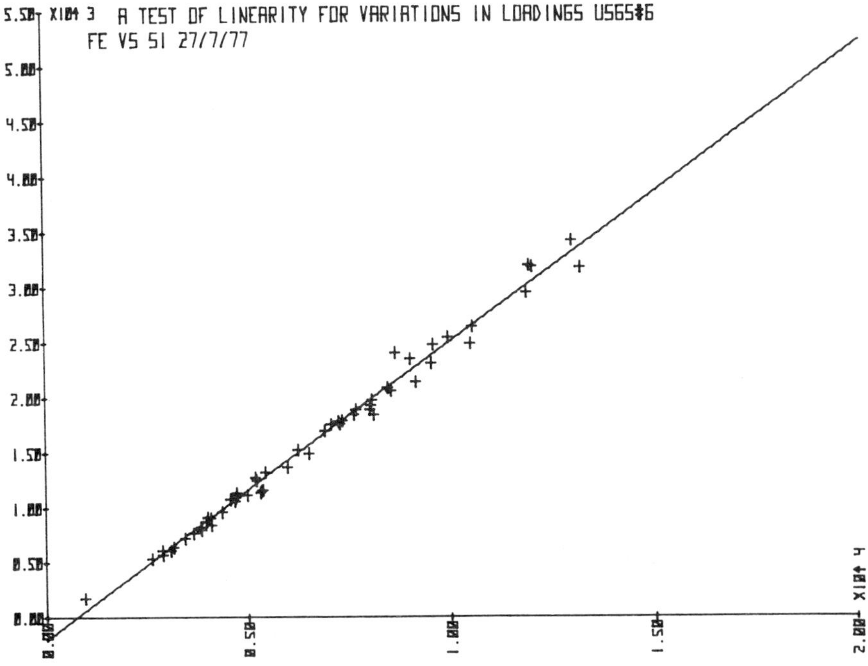

FIGURE V. RATIO PLOT OF IRON AND SILICON RESPONSE FOR VARIED AMOUNTS OF USGS-6 SOIL ON MYLAR TAPE

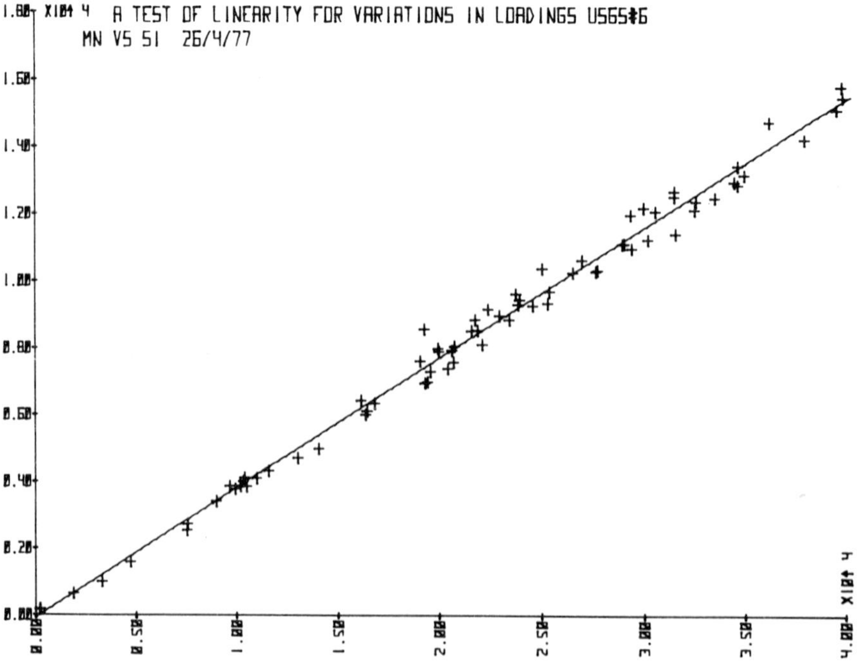

FIGURE VI. RATIO PLOT OF MANGANESE AND SILICON RESPONSE FOR VARIED AMOUNTS OF USGS-6 SOIL ON MYLAR TAPE

Where:

$$RX_j = \frac{ppm_j}{ppm_{Si}} \qquad (6)$$

It is often possible to make reasonable predictions about the most probable chemical composition based upon prior information. For example in a nonurban environment, the following oxides are the most predominant chemical forms encountered; SiO_2, Al_2O_3, Fe_2O_3, CaO, MgO, MnO_2, TiO_2, P_2O_5, etc. (Some judgment is required to determine whether the oxide or the carbonates are the most prevalent forms for calcium and magnesium. However, this can easily be determined by noting the abundance of calcium and magnesium relative to silicon, for the carbonates will generally only be observed when there are low levels of silicon.) It is further reasonable to assume that the oxide form of all the elements monitored comprises 90% of the sample for purposes of normalization. The laser vapourized ppm (LV ppm_j) can be defined as follows:

$$\text{LV ppm}_j = \frac{RX_j \cdot 10^6}{RX_{total}} \, 0.9 \qquad (7)$$

where

$$RX_{total} = SF_{(SiO_2/Si)} RX_{Si} + \sum SF_{(EO_x/E)_j} \cdot RX_j \qquad (8)$$

The validity of the analytical algorithmn for LASERTRACE was evaluated by analyzing geological rock standards comparing the LASERTRACE values to accepted literature values for the content of the standards. Figures 7 and 8 demonstrate the exceptional agreement of LASERTRACE results compared to literature values for beryllium and nickel in 15 different mineral standards comprising 6 USGS rocks, 6

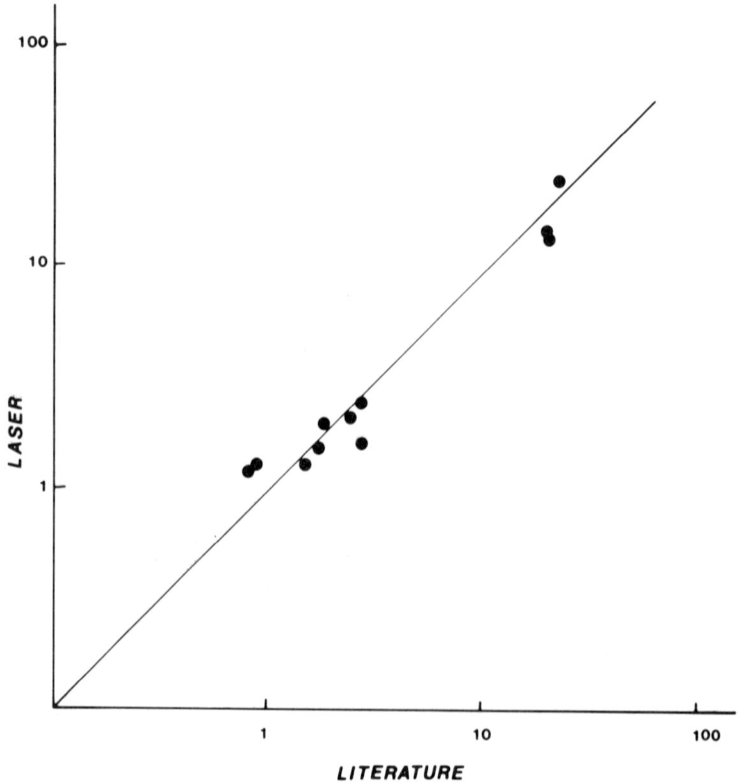

FIGURE VII. COMPARISON OF LASERTRACE ANALYSIS OF Be IN GEOLOGICAL STANDARDS ON MYLAR TAPE WITH LITERATURE VALUES

USGS experimental soil and mineral standards, and 3 CANMET mineral standards. (It was not meaningful to plot all the LASERTRACE results because literature values were not available for these two elements for some of the standards.) Table 2 demonstrates, in tabular format, the agreement obtained for the elements zinc, strontium and copper for 3 USGS rocks.

A comprehensive evaluation of a new analytical technique should demonstrate an ability to analyse "typical real samples." Also, the possibility to use a tape transport system to analyze samples has certain implications with respect to laboratory automation. Suites of soil samples are easily obtained and in most respects soil samples are similar to airborne particulates within certain size fractions. A series of soils collected in a geochemical soil traverse across the

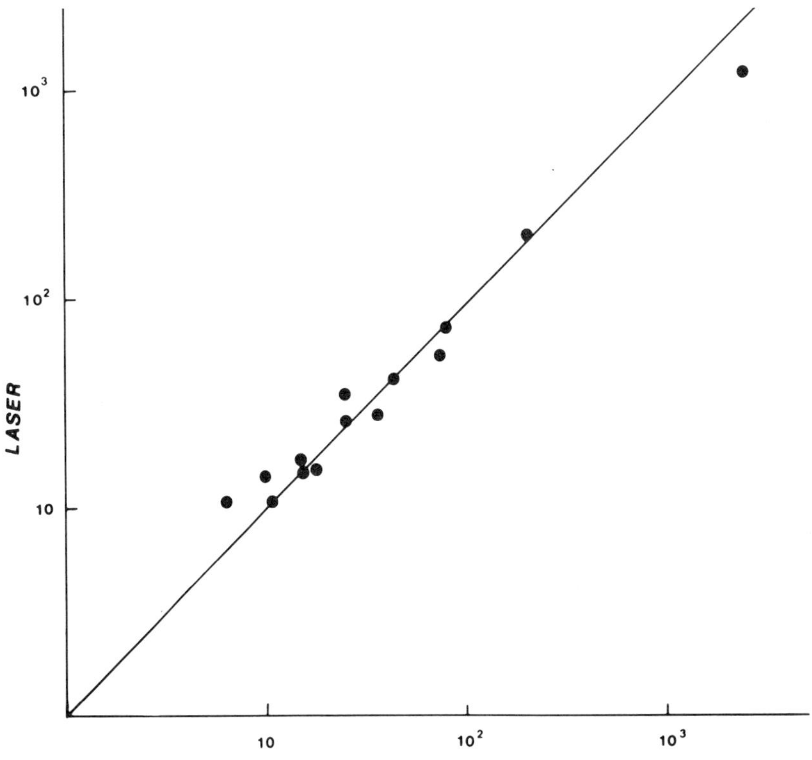

FIGURE VIII. COMPARISON OF LASERTRACE ANALYSIS OF Ni IN GEOLOGICAL STANDARDS ON MYLAR TAPE WITH LITERATURE VALUES

TABLE II. LASERTRACE™ ANALYSIS OF U.S. GEOLOGICAL SURVEY ROCKS.

	Zn ppm		Sr ppm		Cu ppm	
	LASER	LIT	LASER	LIT	LASER	LIT
G-2	83	75	487	463	14	11
GSP-1	128	149	236	247	44	35
AGV-1	76	112	579	657	51	64

Dragoon Oil Field was analyzed by classical acid digestion and by the LASERTRACE approach. (With the LASERTRACE approach, the soils were spread on tape and analysed with the LASERTRACE instrumentation. For the acid digestion approach, the acid leachates from the digestions were introduced to the induction coupled plasma through a pneumatic nebulizer in the classical approach.) Figure 9 compares the results for copper and nickel obtained for the two approaches from the soil traverse samples. Some structural agreement was obtained, but the LASERTRACE approach did not have as clearly defined a geochemical pattern as expected, suggesting that perhaps there was an instrumental matrix problem existing for this sample suite. Microscopic examination of the samples revealed that a significant number of fine particles adhered to the -40 +80 mesh particles in the cases where there was general agreement between the two techniques. There was no correlation between techniques when the fine particles were absent. The -40 +80 mesh size fraction was pulverized, placed on tape and reanalzsed by the LASERTRACE technique yielding better correlation to the wet chemical

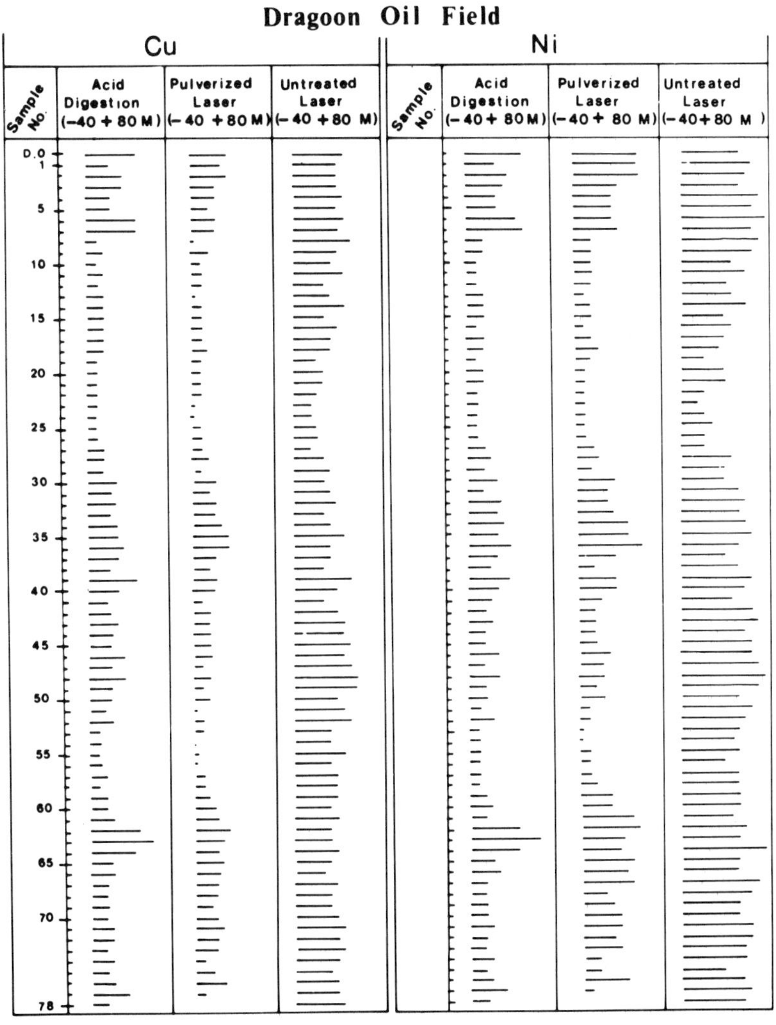

FIGURE IX. COMPARISON OF COPPER AND NICKEL ANALYTICAL DATA OBTAINED BY THE ANALYSIS OF SOIL TRAVERSE SAMPLES

data. These results suggest that differential vaporization was occurring during the LASERTRACE analysis of uncrushed particles. To overcome this differential vaporization we recognised that particles on tape should be pulverized to the 20 - 50 μ range. A sample collection tape deck has been designed in which the samples are automatically pulverized while on the tape using a vibrating anvil striking a stationary anvil. Between samples, the anvils are cleaned by

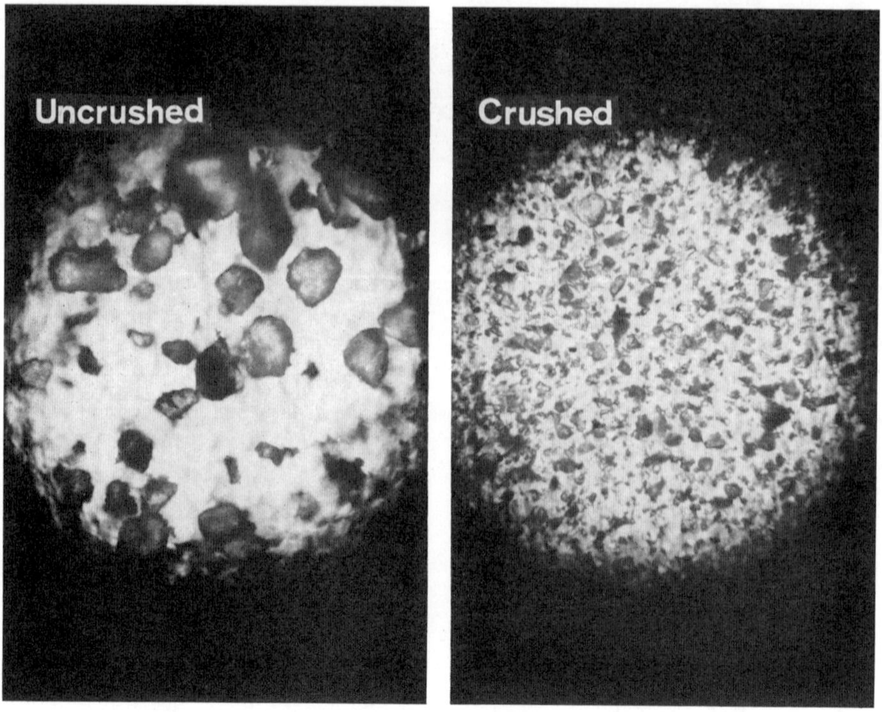

FIGURE X. MICROSCOPIC VIEW OF CRUSHED AND UNCRUSHED SOIL SAMPLES

advancing the tape slightly and vibrating the anvils against fresh clean adhesive tape surface. In Figure 10, the photographs of a microscope view of the "before" and "after" crushing demonstrates the effectiveness of the crusher.

Initially, the requirements of multi-element capability and the ability to detect a variety of elements in a nonurban environment were sited as requirements. These have been achieved. For example, the elements which are often detected in the nonurban environment include: aluminium, calcium, copper, chromium, iron, lead, magnesium, manganese, nickel, strontium, barium, zinc, cadmium, silicon, phosphorous, vanadium, titanium, mercury. (Due to the volatile nature, mercury is monitored "real time" with a sensitive fluorescence spectrometer designed and manufactured by Barringer Research.)

SURVEY RESULTS AND DISCUSSION

Barringer Research has used the AIRTRACE/LASERTRACE instrumentation for geochemical[1] and air quality surveys. The proprietary nature of these studies precludes the release of actual data to the public domain. However, drawing from the experience gained on these surveys, it is possible to construct a conceptual survey which will demonstrate the general type of information that can be obtained from an atmospheric survey. It is possible to conduct such a survey from a van or aircraft, each possessing unique advantages.

The planning stage for a survey includes mapping traverse lines which will intersect the prevailing winds over the survey region. Traverse lines upwind from any source (or at least upwind from the source of interest) provide background level particulate information for control purposes. Downwind traverse lines will include background particulates and pollution particulates additively collected with proper accounting for dilution. (In an actual survey, a variety of grid patterns might be established across the city in an effort to pinpoint pollution sources if they were not already known or to study the dispersal of pollutants as a function of altitude, geographic terrain and meteorological conditions). Figure 11 presents schematically the

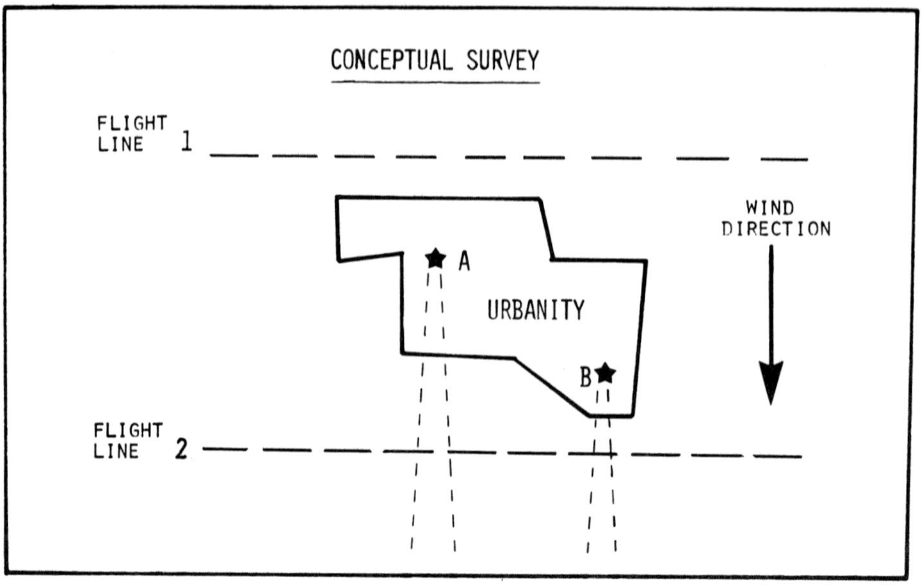

FIGURE XI. SKETCH OF CONCEPTUAL AIR QUALITY SURVEY AREA WITH PROPOSED FLIGHT PATHS

simplest situation where a background (line 1) and monitor (line 2) traverse are employed. In order to simplify the following remarks, vanadium and iron are defined as pollutants while calcium is defined as associated with background sources only. (The choice of vanadium and iron as pollutants has no factual basis in our airborne survey experience.) The hypothetical results are given in Figure 12 for vanadium and reveal a general level associated with the boundaries of the city, a significant burden from source A but no vanadium from source B. (Actual data are often best presented in bar graph format.) The presence of source B can be inferred from the calcium results in Figure 13 where a general "dilution" occurs for the calcium content from

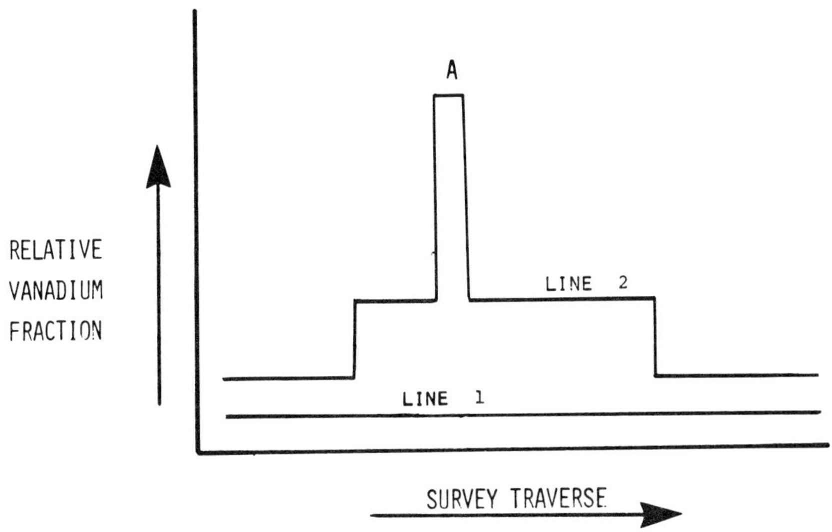

FIGURE XII. RELATIVE VANADIUM CONCENTRATION OF PARTICULATES OBTAINED FROM UPWIND (BACKGROUND) AND DOWNWIND AREAS OF CITY

the background level. If the actual calcium content is known in the background (i.e., via Hi-Vol filters) a semiquantitative indication of the general particulate burden created by the two plumes and the city can be made from the dilution.

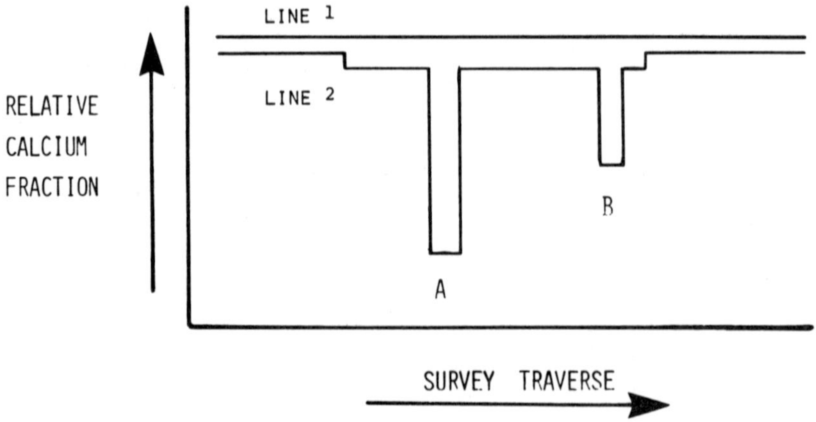

FIGURE XIII. RELATIVE CALCIUM CONCENTRATION OF PARTICULATES OBTAINED FROM UPWIND AND DOWNWIND AREAS OF CITY

The separation of particulates into different size fractions can provide information about the nature of the source or the pollution abatement measures in use at the source. This is represented in Figures 14 and 15 where the iron content of coarse and fine particulates is given. For source A there is relatively little compositional difference between the fine and coarse particulates. However, for source B, the coarse particulates had much less iron content than the fine particulates.

CONCLUSIONS

Barringer Research has been successful in developing analytical instrumentation for the collection and multi-element analysis of air-

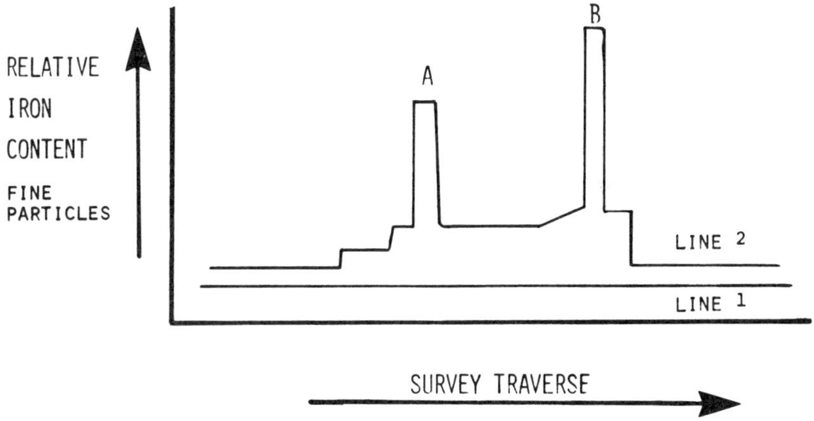

FIGURE XIV. RELATIVE IRON CONTENT OF FINE PARTICULATES OBTAINED FROM UPWIND AND DOWNWIND AREAS OF CITY

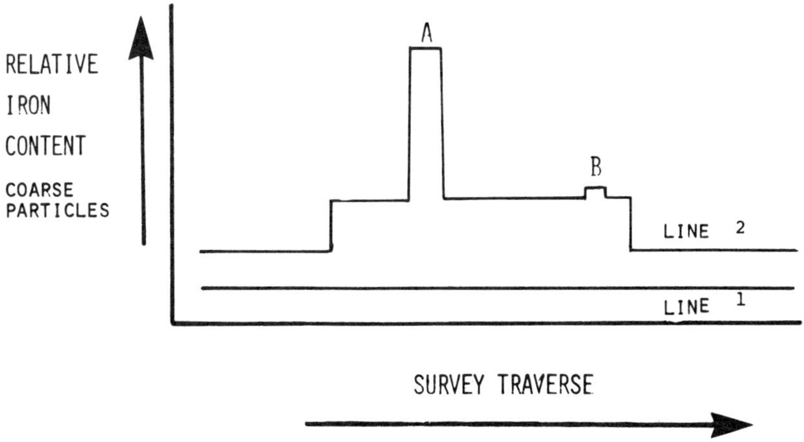

FIGURE XV. RELATIVE IRON CONTENT OF COARSE PARTICULATES OBTAINED FROM UPWIND AND DOWNWIND AREAS OF CITY

borne particulates using AIRTRACE and LASERTRACE. Methodology so developed is sensitive to most common elements at nonurban environment levels, is free from matrix interferences, provides for short sampling intervals and is portable. This system has successfully been used for mineral exploration and air quality surveys.

REFERENCES

1 - A.R. Barringer - "AIRTRACE™ - AN AIRBORNE GEOCHEMICAL EXPLORATION TECHNIQUE," Proceedings of the First Annual William T. Pecora Memorial Symposium, Sioux Falls, S.D., p. 231, October, 1975.

ANALYSIS OF SOIL ANALYSES AND PLANT TISSUE ASH BY ICAP EMISSION SPECTROSCOPY

J. Benton Jones, Jr.

This paper describes the techniques used with an inductively coupled Ar plasma (ICAP) emission spectrometer for the analysis of soil extracts and plant tissue ash. The instrument is located at the University of Georgia in the Institute of Ecology. The ICAP is dedicated primarily to analyses associated with the Institute's research studies on nutrient element cycling in natural and managed environments. Therefore, little research on the plasma system itself is done as the ICP system is fully occupied as a service instrument. Installed in the summer of 1975, the instrument has been analyzing about 1000 to 1500 samples monthly. Initially installed with a modified AA type atomizer (Figure 1) on the plasma torch, a cross-flow atomizer (Figure 2) has been installed recently. The operating characteristics of the ICP system and sample analysis experiences are given in the paper.

Multi-element spark or arc emission spectroscopy has been a technique fairly widely applied to the elemental analysis of plant tissue ash (1-7). Most of the essential plant mineral elements can be determined with relative ease by these emission techniques. The development of plant analysis services for farmers since the early 60's can be traced to the introduction of this multi-element technique providing fast analysis service at a relatively low cost to the user. In addition, much new information on mineral nutrition in plants can be traced to the ability of the researcher to examine a large number of elements easily. More recently, nutrient element cycling studies have been facilitated with these rapid methods of analysis.

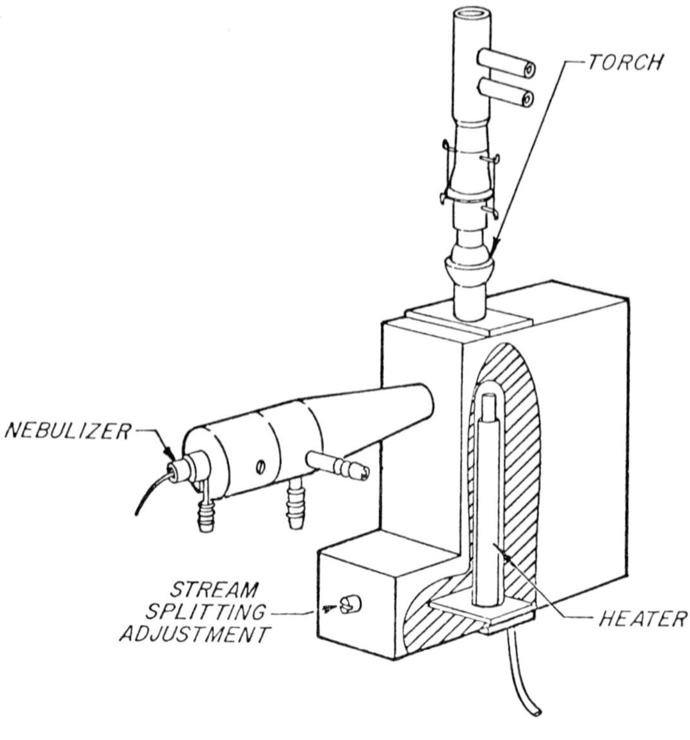

FIGURE I. A.A. NEBULIZER WITH HEATER AND STREAM SPLITTER

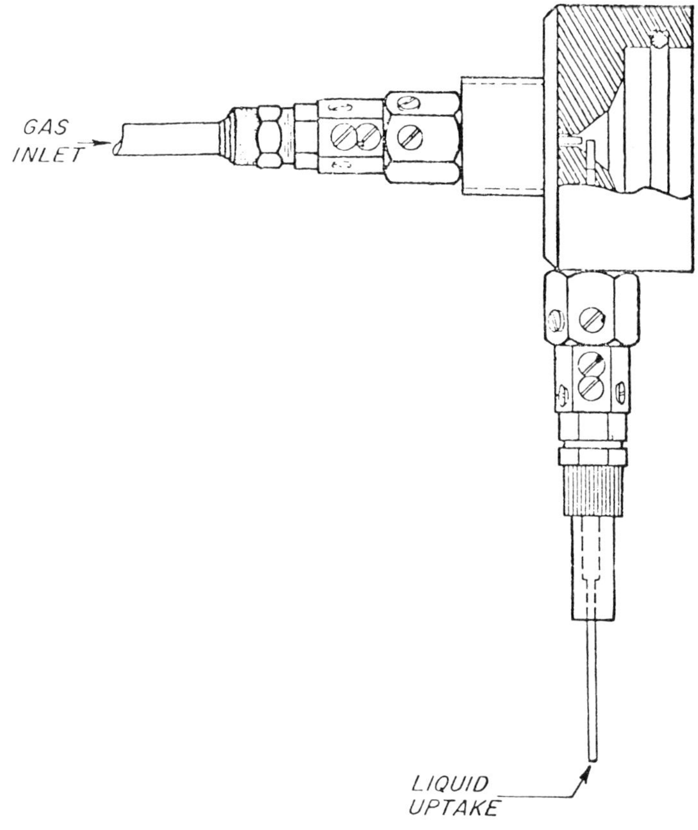

FIGURE II. CROSS FLOW NEBULIZER

As with most analytical techniques, new and improved methods and their resulting instruments are coming into being at a rapid pace. The latest development in emission spectroscopy is the use of plasmas for elemental excitation, replacing the arc or spark as the excitation source. Plasma technology is fairly new and its adaptation to emission spectroscopy quite new. Only in 1974 were the first instruments commercially available. In the last several years, a number of papers have been published on plasma excitation and its analytical application (8-12).

For the biologist, plasma emission spectroscopy offers multi-element analyses of solutions for many elements at sensitivity and precision levels equal to or better than that obtainable using the flame atomic absorption technique. Plasma emission also offers the opportunity to automate sample introduction which is cumbersome or impossible with arc or spark excitation methods.

For the discussion in this paper, the excitation source will be that described by Fassel and Kniseley (8) as the inductively coupled Ar plasma emission technique. An earlier paper by Jones (13) describes the initial experiences with an ICAP emission spectrometer for the analysis of soil extracts and plant tissue digests.

INSTRUMENT CHARACTERISTICS

The ICP system is a Jarrell-Ash Model ICAP AtomComp spectrometer with 19 elements in the program. The spectrometer and plasma operating characteristics are given in Tables 1 and 2.

In order to facilitate sample introduction, an AutoAnalyzer Sampler I is used to introduce samples into the plasma torch. The uptake line is attached to the probe on the Sampler. As the probe moves into the "sample uptake" position, the probe activates the spectrometer, and after an appropriate time delay to allow the nebulized sample to be flowing fully into the torch, the computer begins the integration cycle, records the emission and then activates the teletypewriter to record the results. The analysis pace is set by the AutoAnalyzer Sampler I at 40 samples an hour with a ratio of

TABLE I. SPECTROMETER CHARACTERISTICS

Spectrometer:	Jarrell-Ash 0.75 m, model 750
Fixed Slits:	50-μm entrance 50-μm exit
Grating:	1180 rulings/mm, blazed at 250.0 nm, first order

Element	Spectral Line (nm)	Photomultiplier Tube*
Al	308.2	300
B	249.7	300
Ca	422.6	300
Cd	228.8	427
Co	238.8	427
Cr	357.8	300
Cu	327.4	300
Fe	259.9	300
K	766.4	889
Mg	279.5	300
Mn	257.6	300
Mo	203.8	427
Na	589.0	300
Ni	341.4	300
P	214.9(II)	427
Pb	220.3	427
Pt	265.9	300
Si	288.1	300
Sr	407.7	300
Zn	213.8	427

* Hamamatsu Type "R-"

TABLE II. PLASMA TORCH OPERATING PARAMETERS

Frequency	
Net radio frequency power input to coupling unit	1 kW
Argon coolant flow rate	16 l/min
Argon flow rate for aerosol transport	2 l/min
Aerosol flow rate to plasma	0.5 ml/min

sample to wash as 1 to 2. No operator attention is needed in the hour the 40 samples are analyzed.

An AutoAnalyzer Proportioning Pump III was placed into the sampling system to push the sample solution into the atomizer. The advantage hoped for was to improve the performance of the atomizer by having a constant head pressure on the introduced solution. It has been suspected that the change in viscosity of particularly plant tissue digests does affect the pneumatic action of the cross-flow atomizer. Preliminary tests indicate some reduction in the variance of an analyzed plant tissue standard. A further application of this tested technique of sample introduction is discussed later in the paper. A schematic of the pump tube arrangement is shown in Figure 3.

STANDARDS AND CALIBRATION

Solution standards for the analysis of soil extracts and plant tissue ash are prepared in such a way as to minimize interelement

FIGURE III. AUTOANALYZER FLOW DIAGRAM FOR SAMPLE INTRODUCTION INTO A PLASMA TORCH

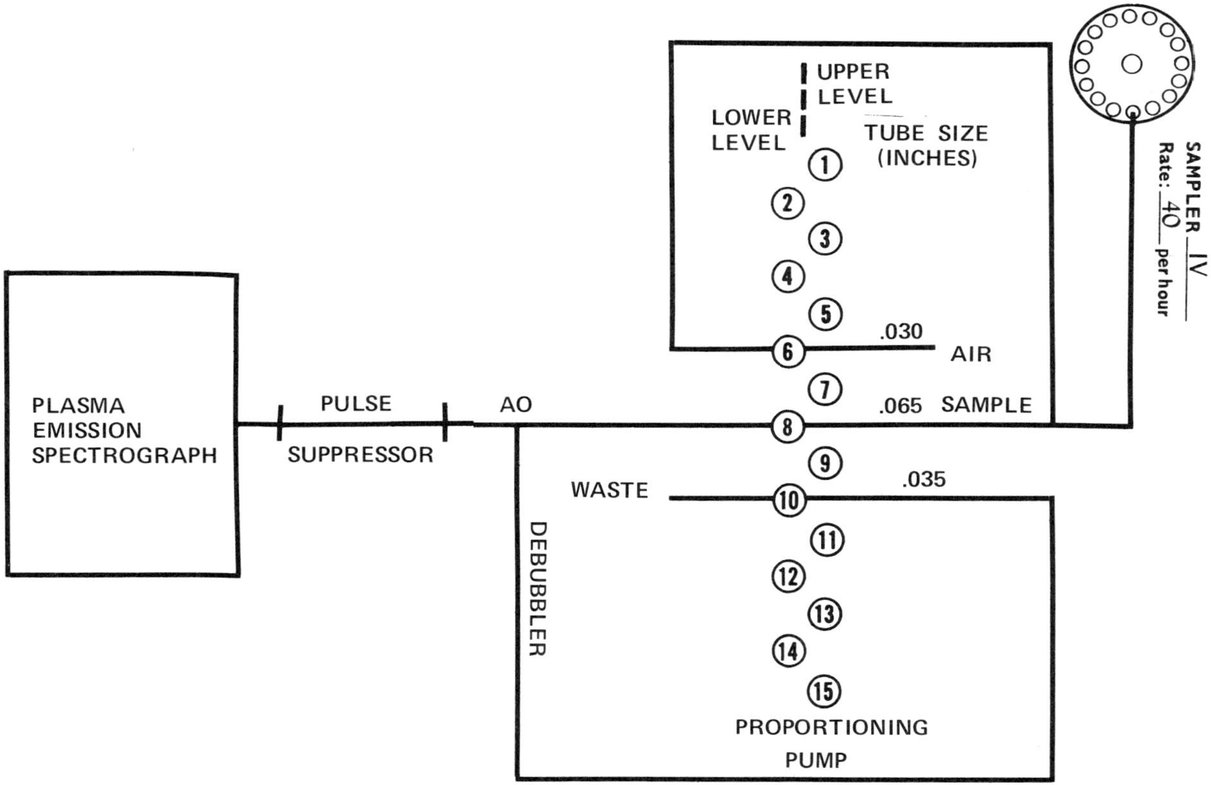

effects and to bracket the anticipated elemental concentration range of the unknowns. After some trial and error experiences, the required standards and their composition were found. It is important that the quality of the prepared standards be carefully controlled, particularly in the selection of reagents due to possible unwanted additions of to-be-determined elements. Good reagent grade chemicals were found satisfactory in most instances. It has become the custom in this laboratory to use 1000 μg/ml atomic absorption elemental standards whenever possible in the preparation of the solution standards. A listing of the standards and their composition is given in Table 3. Deionized water serves as one of the standards. It was found necessary to have an internal standard in the analysis of plant tissue ash solutions and Mo was chosen more for convenience rather than designed choice. Recently, Skogerboe and Coleman (11) found it necessary to use an internal standard. They chose In to minimize what they considered matrix affects in the analysis of food samples by microwave-induced plasma spectroscopy. When the opportunity arises, the plans are to replace Mo with another element to serve as the internal standard.

SAMPLE PREPARATION

(a) Soil Extracts: Nebulizers commonly used on ICP torches are not capable of handling solutions in which the total dissolved solids exceed 1000 μg/ml. Therefore, some commonly used solutions which are concentrated salts, such as 1N neutral ammonium acetate (14), cannot be fed into the nebulizer. However, soil extractants which are dilute

TABLE III. COMPOSITION OF SOLUTION STANDARDS USED TO CALIBRATE THE ICAP SPECTROMETER FOR THE ANALYSIS OF SOIL EXTRACTS AND PLANT TISSUE ASH SOLUTIONS

	Elemental Concentration, μg/ml[1/]				
	Standards for Soil Extracts		Standards for Plant Tissue Ash Solutions		
Element	2	3	MAJH	MINH	CALC
Al	-	10	5	0	0
B	10	0	0	2	0
Ca	10	0	0	0	100[2/]
Cd	0	10	0	2	0
Co	0	10	2	0	0
Cr	10	0	0	2	0
Cu	0	10	5	0	0
Fe	0	10	5	0	0
K	17.50	0	1000[2/]	0	0
Mg	10	0	200[2/]	0	0
Mn	0	10	5	0	0
Mo	0	10	2	0	0
Na	24.50	0	0	0	20
Ni	0	10	2	0	0
P	10	0	20	0	0
Pb	0	10	0	2	0
Pt	10	0	0	0	2
Sr	0	1	0	0	2
Si	10	0	0	5	0
Zn	0	10	0	5	0

[1/] Prepared from 1000 μg/ml stock solutions designated for atomic absorption spectroscopy use.

[2/] Made from 10,000 μg/ml stock solutions.

acids, such as Double Acid (14), are acceptable and can be easily analyzed. Analyses of soil extracts using the Double Acid procedure by the ICAP spectrometer were made and compared to the results obtained by AutoAnalyzer analysis at the Georgia Soil Testing and Plant Analysis Laboratory in Athens. The analyses consisted of four soil extract standards containing known concentrations of Ca, K, Mg and P plus 35 soil extracts. The results of both sets of analyses are given in Table 4.

(b) Plant Tissue Ash: The conventional method of sample preparation is as follows: grind the green dried tissue (at 80°C) to pass a 20 mesh screen and weigh 1.0g into a 15-ml high form porcelain crucible. The sample is dry ashed in a muffle furnace for 4 hours at 500°C. During ashing, the crucible is so placed in the furnace that it is not in physical contact with the heated floor or walls. After cooling, the ash is taken up in 20 ml of 20% HNO_3 containing 10 μg/ml Mo. After allowing the suspended material to settle out, a portion of the cleared solution is transferred to a cup on the AutoAnalyzer Sampler. The sample solution is picked up from the cup and drawn into the atomizer chamber of the plasma torch.

The National Bureau of Standards, Orchard Leaves, Standard Reference Material 1571 was used as a reference to check on the performance of the ICAP spectrometer. Whenever a series of plant tissue ash is analyzed, the orchard leaves standard is run as a check on the calibration. Ten sets in which the orchard leaves sample was included were selected at random covering a three-month period. The mean and

TABLE IV. SOIL SOLUTION STANDARDS AND DOUBLE-ACID SOIL EXTRACTS ANALYSES FOR FOUR ELEMENTS BY ICAP SPECTROSCOPY AND AS REPORTED BY THE GEORGIA SOIL TESTING LABORATORY

	Elemental Content, μg/ml										
	Std. #1		Std. #2		Std. #3		Std. #4		Means		t
Element	Known	Determ	Known	Determ	Known	Determ	Known	Determ	AutoA	Plasma	Value
Ca	15	16	30	31	90	97	150	160	77.1	93.3	2.53*
K	6	7	10	9	30	30	50	50	18.1	17.9	1.41ns
Mg	5	5	15	16	30	31	45	44	16.4	16.1	1.32ns
P	1	1	5	5	20	21	40	43	4.1	4.1	1.01ns

*Statistically significant at the 0.02% level.

nsNot statistically significant.

standard deviation were calculated and the results reported in Table 5. The NBS Certificate of Analysis results are given as well as results reported by Jones (15) in an AOAC study.

TABLE V. ELEMENT CONTENT OF NBS STANDARD ORCHARD LEAVES DETERMINED BY ICAP EMISSION SPECTROSCOPY AND IN A COLLABORATIVE STUDY

Element	Content, $\bar{x}$ ± std. dev.		
	NBS Reported Content[1/]	ICAP Results[2/]	AOAC Results[3/]
	%		
Ca	2.09 ± 0.03	1.99 ± 0.07	1.97 ± 0.20
K	1.47 ± 0.03	1.36 ± 0.07	1.38 ± 0.17
Mg	0.62 ± 0.02	0.62 ± 0.02	0.63 ± 0.04
P	0.21 ± 0.01	0.22 ± 0.005	0.19 ± 0.03
	ug/g		
Al	not reported	234 ± 79	215 ± 49
B	33 ± 3	31 ± 1	32 ± 5
Cd	0.11 ± 0.02	0.41 ± 0.15	not reported
Co	(0.2)	0.23 ± 0.06	not reported
Cr	(2.3)	2.86 ± 0.73	not reported
Cu	12 ± 1	10 ± 1	15 ± 3
Fe	300 ± 20	183 ± 12	241 ± 62
Mn	91 ± 4	101 ± 2	94 ± 27
Na	82 ± 6	76 ± 10	206 ± 183
Pb	(44)	40 ± 3	not reported
Si	not reported	437 ± 46	not reported
Sr	(37)	34 ± 0.6	29 ± 8
Zn	25 ± 3	26 ± 1	28 ± 6

[1/]National Bureau of Standards, Certificate of Analysis, Standard Reference Material 1571, revised August 15, 1976.

[2/]Based on 10 different days results over a 2 month period in 1977.

[3/]Reported in JAOAC paper (6).

RESULTS AND DISCUSSION

One measure of performance for an ICAP system is the determined detection limit for each programed element. Detection limits vary from day to day, even hour to hour, although the variation is usually in the second significant digit and not at the decade level. Typical detection limits for the ICAP instrument described in this paper are given in Table 6. Detection limits reported by Fassel and Kniseley (8) are also given for comparison. However, it should be remembered that the Fassel-Kniseley detection limits were reported based on the best of operating conditions while the detection limits for the spectrometer being discussed here are for an instrument engaged in daily routine analyses where all conditions of operation for every element are not ideally set. However, the detection limits reported here are adequate for most elements found in soil extracts and plant tissue ash solutions.

To test the feasibility of using the ICAP spectrometer for routine soil extract analyses, a tray of 35 Double Acid soil extracts provided by the Georgia Soil Testing and Plant Analysis Laboratory were analyzed and the results compared with those reported by the soil lab (see Table 4). Results for K, Mg and P were found to be not statistically different by the two methods of analysis; however, Ca was. Analysis of the standards used to calibrate the AutoAnalyzer in the soil lab revealed that the calibration routine in the ICAP system did not agree with the AutoAnalyzer standards. This discrepancy accounts for the difference in results. No attempt was made to uncover the calibration error for this set of results as we are able to successfully demonstrate that routine soil analyses can be easily handled by the ICAP system with excellent results.

TABLE VI. ELEMENT CONCENTRATION IN TISSUE DIGESTS COMPARED TO ICAP DETECTION LIMITS

	μg /ml		
	Concentration 1 g ash to 20 ml		
Element	Orchard Leaves	Bovine Liver	ICAP Detection Limits
Ca	1045	6.15	0.025
K	735	485	3.400
Mg	310	30.25	0.001
P	105	--	0.130
B	1.65	--	0.170
Cu	0.60	9.65	0.022
Fe	15.00	13.50	0.013
Mn	4.55	0.52	0.031
Zn	1.25	6.50	0.003
Cd	0.005	0.013	0.0024
Co	0.010	0.009	0.0023
Cr	0.15	--	0.025
Ni	0.065	--	0.006
Pb	2.250	0.017	0.047
Sr	1.85	0.007	0.045

It is expected that other elements could also be determined as usable concentration results were recorded for the elements Al, Cu, Fe, Mn, Na, Si and Zn.

Soltanpour and Schwab (16) recently reported on a new soil extraction procedure using a modified Olson extractant (14). An ICAP spectrometer was used to assay the extracted elements, Cu, Fe, K, P, Mn and Zn, with good results. This procedure is now in use in the Colorado Soil Testing Laboratory.

The ICAP technique is also ideally suited for the analysis of extracts prepared by the new Baker (17) method of soil analysis where a single procedure can be used to evaluate the status of 11 elements, Al, Ca, Cd, Cu, Fe, K, Mg, Mn, Na, Ni and Zn, in the soil.

The speed and multi-element capacity of ICAP instruments makes them ideally suited for use in soil testing laboratories when both major and micronutrient elements are determined in the same soil extract. The use of water, supposedly the "ideal" extractant, may be possible with an ICAP spectrometer doing the assay where sub parts-per-million sensitivity would be needed.

Repeated analyses of an ashed sample of the NBS Reference Standard Orchard Leaves, 1571, were used to evaluate the ICAP system for the analysis of plant tissue ash (see Table 5). The means are comparable except for the results reported for the elements Cd and Fe. The high Cd results are due to a Ca background interference and inadequate background correction. A careful evaluation of the background problem and its correction are needed. The low Fe results are due to method of sample preparation. Dry ashing gives low results. Other than these element discrepancies the comparison of results for the orchard leaves standard is acceptable for routine mass analysis of large numbers of plant tissue ash.

AUTOMATED PLANT DIGEST ANALYSIS

As described earlier in this paper, sample pick-up is by the use of an AutoAnalyzer Sampler I. Once the ICP system is calibrated, 40 samples are analyzed without operator assistance.

The next step was to add an AutoAnalyzer Proportioning Pump III to the system which could perform the functions of diluting the prepared plant tissue digest and adding of an internal standard. But primarily, the pumping of the sample into the nebulizer under pressure was thought to improve the nebulizer performance. Performance data using a reference plant tissue digest does show a reduction in percent relative standard deviation for all elements except Cd and Zn. Additional experience and experimentation are needed, however, to more thoroughly evaluate this method of sample introduction.

The most significant application for the automation of sample introduction may come by further refinement of the sample itself. In the analysis of plant tissue digests, the major elements, such as Ca, K, Mg and P, are present in concentrations in hundreds to thousands of milligrams per liter, while the micronutrients and other trace elements are in few or sub milligrams per liter. Therefore, although accuracy in determination is essential for all elements, lack of adequate sensitivity is of no concern for many of the essential elements with plasma excitation. However, for most of the trace elements, sensitivity is a major concern.

The current procedure is to dry ash (at 500°C for 4 hours) a 1-gm oven dried (80°C) sample and to take the ash up to a volume of 20 ml with 20% HNO_3 containing 10 μg/ml Mo which serves as the internal standard. Based on this dilution, the elemental concentration in the digest for the major, micronutrient, and trace elements for the NBS Orchard Leaves and Bovine Liver Standards compared to the elemental sensitivity for plasma excitation is shown in Table 6. The concentration of the major elements, Ca, K, Mg and P in the to-be-analyzed solution ranges from 1000 to 1,000,000 times greater than the detection limits. For the trace elements, Cd, Co, Cr, Ni and Pb, the detections limits are fairly close to the concentrations present in the tissue ash solution.

There has been much in the literature regarding the troublesome effect of Ca on the emission characteristics for some of the trace elements, as Ca in high concentrations presents a considerable background problem (18). The ICP system may be programmed to compensate or minimize the Ca emission effects for all elements with particular attention for those elements in trace concentrations and of biological interest. However, these compensating techniques may be of limited success. An ideal solution would be to remove Ca from the ash solution as well as other major elements which may give similar interfering effects. An

AutoAnalyzaer system may offer just such a possibility.

It is conceivable that a 1-gm plant tissue sample can be homogenized in an acid digestion mixture of HNO_3-$HClO_4$ acids, digested in the AutoAnalyzer Helix Digestor, divided into two analytical streams, the one to be pumped directly into the nebulizer for major element analysis and the other through the system which selectively removes the major elements prior to entering the nebulizer. A schematic drawing of the system is shown in Figure 4. In a similar manner, the ash solution or even a HOAc extraction of a plant tissue sample can be handled in much the same manner. This latter technique may be particularly useful since the ICAP has excellent sensitivity.

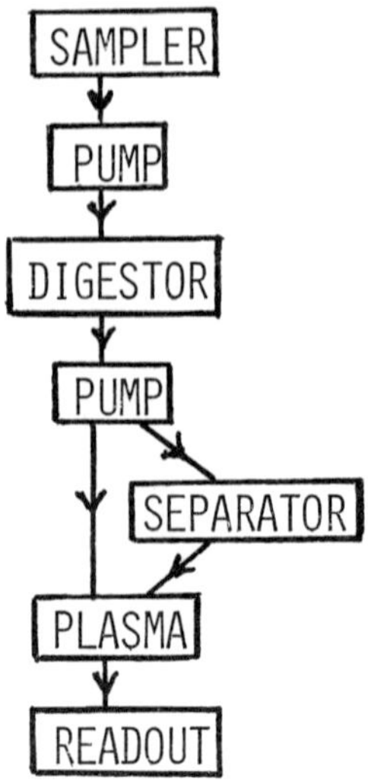

FIGURE IV. SAMPLE FLOW DIAGRAM FOR PREPARING PLANT TISSUE ASH FOR MACRO- AND MICRO-ELEMENTS BY ICAP EMISSION SPECTROSCOPY

The ultimate goal of the Technicon AutoAnalyzer preparation unit for feeding digested samples into a ICAP spectrometer is to achieve a high performance analytical system both in number of samples analyzed and number of elements determined irrespective of concentration with minimum labor input. The Sample I and Proportioning Pump III components have been tested and were found to be functionally successful.

The next step is to determine if improved performance and sensitivity can be gained for the trace elements in plant tissue digests when Ca and the other major elements are removed. If so, then an automated digestion and separation system would be worthy of further investigation.

SUMMARY

The analysis system described in this paper has been successfully in operation for over two years. The method of sample preparation and analysis gives excellent results. There are needed improvements in technique and they have been discussed.

REFERENCES

(1) D. E. Baker, et al., Agron. J. 50:133 (1964).

(2) P. N. Carpenter, et al., Maine Agr. Exp. Sta. Tech. Bul. 30 (1968).

(3) R. E. Christensen, R. M. Beshman, and J. J. Birdsall, J. Assoc. Offic. Anal. Chem., 51:1003 (1968).

(4) J. B. Jones, Jr., Amer. Lab. 8:15 (1976).

(5) A. L Kenworthy, Proc. 36th Ann. Mtg., Council on Fert. Appl., NPFI, Washington, D. C., pp. 36-50 (1960).

(6) R. L. Mitchell, Tech. Comm. No. 44, Commonwealth Bureau of Soils, Harpenden, Hert., England (1950).

(7) G. Thompson and D. Bankston, Spectrochim. Acta, 24B:335 (1969).

(8) V. A. Fassel and R. N. Kniseley, Anal. Chem. 46:1110A (1974).

(9) V. A. Fassel and R. N. Kniseley, Ibid., 46:1155A (1974).

(10) R. K. Skogerboe, I. T. Urasa, and G. N. Coleman, Appl. Spectrosc. 30:500 (1976).

(11) R. K. Skogerboe and G. N. Coleman, Ibid., 30:504 (1976).

(12) D. K. Christopher and T. J. Vickers, Ibid., 31:273 (1977).

(13) J. B. Jones, Jr., Comm. Soil Sci. Plant Anal., 8:349 (1977).

(14) J. B. Jones, Jr., (ed.), Handbook on Reference Methods for Soil Testing, Coun. Soil Test. Plant Anal., Athens, Ga. (1974).

(15) J. B. Jones, Jr., J. Assoc. Offic. Anal. Chem., 58:764 (1975).

(16) P. N. Soltanpour and A. P. Schwab, Agron. Abstr., Amer. Soc. Agron., 1977:166 (1977).

(17) D. E. Baker, Soil Sci. Soc. Amer. Proc. 37:537 (1973).

(18) D. K. Christopher and T. J. Vickers, Appl. Spectrosc., 31:275 (1977).

TRACE METALS ANALYSIS IN SILICON AND ALUMINUM METALS BY INDUCTIVELY COUPLED PLASMA

W. Zamechek, R.J. Lewandowski,
R.G. Parkhurst, and A.J. Ellgren

As a part of a continuing program in trace metal methodology, procedures have been developed in our laboratory for the determination of trace elements in Si and Al metals, as well as their oxides, by Inductively Coupled RF Plasma Emission Spectroscopy.

The instrumentation and operating parameters used for this work are given in Table I. The system consists of Model 975 Jarrell-Ash Plasma atomcomp spectrometer with a Plasma-Therm 2 kW generator @ 27.13 MHz. The spectrometer is equipped with an automatic background corrector. The program elements, their wavelengths, and the lowest quantitatively determined concentrations (LQDC's) are listed in Table II. LQDC is defined as being 5 times the detection limit. The precision, expressed as percent relative standard deviation at the 1.0 and 0.1-µg/ml concentration level, is also shown for an aqueous matrix.

Since liquid samples are required for ICP analysis, various sample decomposition methods have been developed for the matrices studied. J. T. Baker Ultrex grade acids are used throughout.

TABLE I. OPERATING SYSTEM

Instrumentation	JARRELL-ASH PLASMA ATOMCOMP Model 975 equipped with an automatic background corrector 2kW Plasma-Therm RF generator @ 27.13 MHz
Nebulizer	JARRELL-ASH cross flow
Gas	Coolant Ar at 18 L/min Sample Ar at 1.2 L/min Plasma Ar at 0 L/min
Sample Uptake Rate	≃ 1 mL/min
Observation Zone	20 mm above coil
Exposure Times	10 sec without the background corrector 4 sec/cycle with the background corrector
Torch	Jarrell-Ash, inner tube diameter 1.5 mm

TABLE II. PROGRAM ELEMENTS AND WAVELENGTHS

LQDC's*and % RSD in an Aqueous Matrix**

Element	λ (nm)	LQDC μg/ml	% RSD 1 μg/ml	0.1 μg/ml
Ag	328.0	0.071	0.47	0.50
Al	308.2	0.045	0.48	3.75
As	193.6	0.100	2.25	14.30
Au	242.7	0.015	0.60	2.08
B	249.8	0.009	0.65	2.84
Ba	455.4	0.002	0.65	0.62
Be	313.0	0.001	0.63	0.64
Ca	396.8	0.002	0.58	0.80
Cd	226.5	0.003	0.46	0.62
Ce	446.0	0.260	4.00	--

* LQDC = Five times the detection limit

** No background corrector

TABLE II. (Continued)

LQDC's* and % RSD in an Aqueous Matrix**

	λ	LQDC	% RSD	
Element	(nm)	μg/ml	1 μg/ml	0.1 μg/ml
Co	238.8	0.010	0.44	0.61
Cr	357.8	0.090	0.99	30.00
Cu	324.7	0.008	0.49	0.88
Fe	259.9	0.010	0.51	0.65
Ga	417.2	0.084	0.61	11.00
La	379.4	0.015	0.60	1.23
Li	670.7	0.028	0.63	1.27
Mg	279.5	0.002	0.47	0.85
Mn	257.6	0.008	0.49	1.14
Mo	202.0	0.120	3.66	14.00
Na	589.0	0.600	2.99	21.00
Nb	318.0	0.090	1.49	14.00
Ni	354.4	0.082	2.99	4.91
P	213.6	0.180	1.67	46.00
Pb	220.3	0.076	0.63	2.98
Pd	363.4	0.390	3.97	--
Pt	265.9	0.068	1.10	15.00
Rh	343.4	0.087	1.24	8.88
Sb	206.8	0.061	2.17	5.98
Se	196.0	0.085	2.56	27.00
Si	288.1	0.033	0.60	6.10
Sn	189.9	0.030	1.01	3.59
Sr	421.5	0.001	0.63	0.64
Ti	334.9	0.004	0.60	0.55
V	310.2	0.042	0.73	3.39
W	202.9(2)***	--	--	--
Zn	206.1(2)***	0.005	0.53	1.27
Zr	339.1	0.035	0.95	1.95

* LQDC = Five times the detection limit

** No background corrector

*** Second order wavelength

Silicon metal and silica are suitable for acid decomposition with HF. The silicon matrix can be removed by volatilization with HF. The decomposition procedure for Si metal is given in Table III, and the procedure for silicon dioxide in Table IV. These procedures can be varied by altering the sample size to reach the desired detection limits in the solid samples. Any insoluble residues present following these dissolution procedures can be treated separately by further chemical methods. Volatilization of the matrix results in what is essentially a pure aqueous matrix, thus no background correction is necessary. LQDC's and precision obtained are the same as indicated in Table II. A Spex silicon metal standard containing trace metals at the 0.01% concentration level was analyzed by the above procedure. Sample dissolution resulted

TABLE III. Si DECOMPOSITION PROCEDURE

1. Weigh 1 g. metal into a 25 ml Teflon PFA screw cap vial.
2. Add 15 ml HF.
3. Add HNO_3 dropwise until all Si is in solution.
4. Add 3 drops $HClO_4$.
5. Evaporate with IR heater to dryness.
6. Add 0.6 ml HCl and 1 ml H_2O.
7. Seal and heat gently for about 30 minutes.
8. Add 9 ml H_2O.
9. Seal and heat for 1 hour at 105^0C.
10. Cool, analyze.

TABLE IV. SiO_2 DECOMPOSITION PROCEDURE

1. Weigh 1 g. of powdered sample into a Parr Teflon lined bomb.
2. Add 15 ml HF and 5 ml HCl.
3. Close bomb and heat at 120°C for a minimum of 2 hours.
4. Cool and transfer into a Teflon beaker.
5. Evaporate to dryness with an IR heater.
6. Add 2 cc. of 1 - 1 HCl.
7. Heat gently for 30 minutes.
8. Transfer into a polypropylene 10 ml volumetric flask.
9. Dilute to mark, analyze.

in 0.31 μg/ml analyte in solution. The values for several elements in solution were checked by atomic absorption using flame and flameless methods. Cr is present in the Spex standard as Cr_2O_3 and is not completely dissolved by this dissolution technique. The results are presented in Table V. The values obtained by ICP and atomic absorption compared well.

Aluminum metal is decomposed with hydrochloric acid. The procedure is outlined in Table VI. Aluminum oxides are decomposed with hydrochloric acid in a Teflon-lined bomb, the decomposition conditions are listed in Table VII. The conditions are dependent on the type of alumina present. Reaction conditions range from 2 hours at 150°C for γ-alumina, to 24 hours at 240°C for α-alumina and sapphire. Particle size for the refractory forms of alumina should be -200 mesh. Grinding of sapphire is

TABLE V. ANALYSIS OF SPEX 0.01% ELEMENTS IN Si METAL STANDARD. (0.31 μg/ml Analyte Concentration)

	% Recovery	
Elements	ICP	AAS
Al	60	58
As	100	--
Ba	95	--
Be	100	--
Ca	101	--
Cd	99	--
Ce	103	--
Co	100	--
Cu	130	128
Fe	138	139
Ga	80	--
Li	103	--
Mg	106	105
Mn	102	--
Mo	95	--
Ni	120	126
P	130	--
Pb	120	116
Sr	100	--
Ti	97	--
V	103	--
Zn	109	108
Zr	110	--

especially difficult. We use a high purity sapphire ball and vial for grinding of sapphire samples. Some of the decomposition conditions have been supplied by Prof. Franz Rozenberger of Utah University who is involved with sapphire analysis. Again, any insoluble residue present after the above procedure can be treated separately.

TABLE VI. Al DECOMPOSITION PROCEDURE

1. Weigh 0.25 g. sample* (piece, powder or chip) into a 100-ml beaker.
2. Add 16 ml 25% HCl.
3. After dissolution transfer into a 50-ml volumetric flask.
4. Dilute to mark, analyze.

*Prior to dissolution the surface of the sample may have to be cleaned with dilute HCl.

TABLE VII. α Al_2O_3 DECOMPOSITION PROCEDURE

1. Weigh 0.2 g. sample powder (-200 mesh) into a Teflon-lined, high pressure Parr bomb.
2. Add 4 ml HCl and 0.5 - 1 ml H_2O.
3. Close bomb and heat @ 240°C for 24 hours.
4. Cool and transfer into 25-ml volumetric flask, dilute to mark, analyze.

If sample is δ Al_2O_3 or Boehmite, bomb @ 150°C for 2 hours.

TABLE VIII. LQDC AS A FUNCTION OF Al CONCENTRATION

HCl Concentration at ≃ 10% Background Corrector D = 1

Element	0.46% Al µg/ml	0.30% Al µg/ml	0.15% Al µg/ml	0.05% Al µg/ml	0% Al µg/ml
As	5.400	2.700	1.400	0.620	0.300
Au	0.035	0.024	0.027	0.020	0.019
B	0.089	0.088	0.095	0.090	0.100
Ba	0.007	0.007	0.009	0.008	0.006
Be	0.001	0.001	0.001	0.002	0.001
Ca	0.009	0.009	0.011	0.010	0.010
Cd	0.027	0.020	0.018	0.012	0.010
Ce	0.800	0.630	0.640	0.800	0.760
Co	0.016	0.015	0.016	0.017	0.015
Cr	0.180	0.110	0.110	0.100	0.080
Cu	0.016	0.013	0.016	0.013	0.013
Fe	0.018	0.015	0.012	0.010	0.008
Ga	0.100	0.120	0.120	0.100	0.110
La	0.018	0.014	0.015	0.013	0.010
Li	0.033	0.031	0.030	0.035	0.031
Mg	0.002	0.002	0.002	0.002	0.002
Mn	0.023	0.021	0.026	0.025	0.024
Mo	0.260	0.180	0.081	0.080	0.077
Na	2.400	1.800	1.000	1.100	1.200
Nb	0.120	0.160	0.100	0.100	0.100
Ni	0.110	0.120	0.130	0.120	0.120
P	1.600	0.830	0.380	0.450	0.420
Pb	0.700	0.350	0.120	0.120	0.130
Pd	0.400	0.400	0.390	0.350	0.340
Pt	0.500	0.130	0.150	0.100	0.080
Rh	0.130	0.150	0.160	0.100	0.080
Sb	1.300	0.580	0.290	0.140	0.130
Se	0.970	0.720	0.410	0.140	0.160

TABLE VIII. (Continued)

HCl Concentration at ≃ 10% Background Corrector D = 1

Element	0.46% Al μg/ml	0.30% Al μg/ml	0.15% Al μg/ml	0.05% Al μg/ml	0% Al μg/ml
Si	0.084	0.062	0.061	0.058	0.055
Sn	0.320	0.250	0.080	0.060	0.050
Sr	0.001	0.001	0.001	0.001	0.001
Ti	0.006	0.006	0.006	0.004	0.003
V	0.160	0.088	0.100	0.090	0.086
Zn	0.094	0.062	0.052	0.019	0.018
Zr	0.056	0.030	0.024	0.016	0.011

We work at a maximum Al concentration of about 0.5 wt %. Higher matrix concentration results in excessive drift due to build-up of solids on the torch tip. Higher matrix concentrations can be used with more frequent restandardization and torch cleaning. Since 0.5% Al produces stray light and a recombination continuum which affect a large number of elements, especially those whose analytical wavelengths are lower than 240 nm, a background corrector was utilized for aluminum work. The interferences are most severe on As, Mo, P, Pb, Sb, Se, and Sn. If these elements are not of interest, the background corrector does not have to be used. The sensitivity obtained with the background corrector is poorer, but the long-term stability is better, due to automatic measurement of line and background intensities with computerized correction. The background corrector, also known as a spectral shifter, examines the line intensity and the background intensity on either, but not both sides of the line. Exposure times of 4 seconds are used for each measurement, the minimum cycle being an average of 2 measurements of the line and one of the background intensities; this is represented as D = 1. At D = 2 the background is measured twice and the line three times, and so on. The averaged background measurements are then subtracted from the averaged line intensity measurements automatically by the computer. The LQDC's are listed in Table VIII as a function of matrix concentration; the HCl concentration was held constant at approximately 10% HCl. As expected the LQDC's for the elements most effected by the matrix, for example As, Sb, and Cd, improved as the Al concentration decreased. The percent recoveries as a function of the matrix concentration are listed in Table IX. All elements were present at 10 μg/ml in the aluminum matrices. HCl concentration in 0.46% aluminum was at approximately 10% and proportionally lower for the other matrix concentrations. The unusual behavior of Li and especially Nb has not been investigated, although we suspect that in the case of Li the aluminum matrix supresses the ionization of Li, thus enhancing atomic emissions. In general these data show the expected trend; i.e., a decrease in percent recovery with an increase in matrix concentration. The magnitude of the supression is larger than can be expected from a 0.5% solids matrix, and is due to the relatively high concentration of hydrochloric acid, especially in the 0.46% Al matrix. Other studies done in our laboratory and published data [1] clearly demonstrate the

---

[1] S. Greenfield, H. McD. McGeachin, and F. A. Chambers, <u>ICP Information Newsletter</u>, 3, 117-127 (1977).

TABLE IX. % RECOVERY AS A FUNCTION OF Al CONCENTRATION. All elements at 10 μg/ml. Background corrector D = 1.

Element	0.46% Al 10% HCl	0.30% Al 6% HCl	0.15% Al 3% HCl	0.05% Al 1% HCl
As	92	95	98	104
Au	87	95	96	99
B	82	88	94	97
Ba	87	91	94	101
Be	93	98	99	100
Ca	93	96	98	100
Cd	86	89	93	100
Ce	73	79	88	95
Co	88	91	95	101
Cr	84	89	95	98
Cu	89	92	95	99
Fe	86	90	94	99
Ga	84	91	92	97
La	85	91	94	97
Li	134	126	117	109
Mg	86	88	91	97
Mn	86	89	93	98
Mo	88	91	94	97
Na	109	105	104	98
Nb	122	115	87	70
Ni	89	91	93	97
P	85	91	96	99
Pb	82	86	92	97
Pd	84	90	92	98
Pt	87	92	91	97
Rh	80	88	91	96
Sb	87	90	94	99
Se	85	87	94	100
Si	85	88	94	97
Sn	86	90	93	98
Sr	90	92	96	99
Ti	85	90	93	97
V	86	89	93	97
Zn	88	91	95	98
Zr	86	92	98	101

TABLE X. PERFORMANCE IN 0.46% Al. Background corrector D = 2.

Element	LQDC µg/ml	% RSD @ 5 µg/ml	% RSD @ 1 µg/ml	% RSD @ 0.1 µg/ml
As	3.500	3.980	19.000	--
Au	0.002	0.690	1.040	1.870
B	0.057	0.630	2.880	8.570
Ba	0.006	1.260	2.140	2.560
Be	0.001	0.980	1.510	1.860
Ca	0.130	0.690	0.950	1.920
Cd	0.024	0.840	1.180	3.400
Ce	0.800	1.800	13.000	--
Co	0.013	0.790	1.010	1.740
Cr	0.100	0.980	4.880	13.000
Cu	0.022	0.600	0.940	1.630
Fe	0.013	0.810	2.370	2.720
Ga	0.100	0.930	2.000	20.000
La	0.015	0.860	1.240	2.240
Li	0.033	1.670	3.330	3.830
Mg	0.002	0.500	1.140	1.630
Mn	0.016	0.450	1.130	2.450
Mo	0.190	1.220	3.850	42.000
Na	1.280	2.670	7.590	--
Nb	0.090	1.000	1.620	5.000
Ni	0.130	1.250	2.980	23.000
P	1.000	1.450	9.700	--
Pb	0.400	2.280	4.390	--
Pd	0.440	1.580	4.580	--
Pt	0.220	1.640	3.330	23.000
Rh	0.066	0.980	1.340	18.000
Sb	0.770	2.330	15.000	--
Se	0.680	1.940	7.640	--
Si	0.047	0.600	2.220	14.000
Sn	0.200	1.260	5.290	66.000
Sr	0.001	1.040	2.090	2.260
Ti	0.010	1.180	3.380	11.000
V	0.110	1.640	3.330	10.000
Zn	0.044	0.670	1.280	5.820
Zr	0.130	0.910	2.580	2.700

suppression effects of mineral acids at high concentrations. The suppression is caused by differences in nebulizer performance in aqueous and acid matrices. The data in Table IX demonstrate the importance of matching the acid content as well as the matrices of samples and standards. Internal standards are commonly used in emission spectroscopy when utilizing arc or spark sources. Internal standards can also be used with an RF plasma source. The data in Table IX also demonstrate the importance of choosing the proper internal standard. For example, Li would not be a good internal standard for the determination of any of the other elements in the aluminum matrix. We did not use an internal standard for this work.

The precision obtained in 0.46% Al at analyte concentrations of 5.0, 1.0, and 0.1-μg/ml is given in Table X.

Comparison of the analysis of actual sapphire samples with ICP, atomic absorption, and semiquantitative emission spectroscopy are listed in Table XI, and the results obtained for the analysis of Al reference standards are listed in Table XII. Excellent agreement between the ICP and reference values has been obtained.

In conclusion, procedures have been developed in our laboratory for the determination of trace metals in silicon and aluminum metals and in their oxides by ICP spectroscopy. Excellent agreement has been found between the ICP and reference values. Detection limits in the low part-per-million range can be obtained for a large number of trace metals. The decomposition procedures are simple and require minimal operator time and sample handling.

TABLE XI. SAPPHIRE ANALYSIS - COMPARISON DATA

	Sample 1			Sample 2		
Element	μg/g ICP	μg/g AAS	μg/g Emission Spec Semi-Quant.	μg/g ICP	μg/g AAS	μg/g Emission Spec Semi-Quant.
As	<250	--	<100	<250	--	<100
Au	<4	--	--	<4	--	--
B	<4	--	<5	<4	--	<5
Ba	18	--	<200	9	--	<200
Be	<4	--	<3	<4	--	<3
Ca	84	--	50 - 150	60	--	40 - 140
Cd	<8	--	<50	<8	--	<50
Co	<4	--	<10	<4	--	<10
Cr	70	61	30 - 60	43	40	20 - 50
Cu	4	2	5 - 12	<4	<2	<5
Fe	386	375	200 - 350	189	176	125 - 225
Ga	41	--	30 - 50	13	--	<30
La	<4	--	--	<4	--	--
Li	<4	--	--	<4	--	--
Mg	37	--	20 - 60	1639	--	1000 - 3500
Mn	234	250	100 - 250	171	169	100 - 250
Mo	<10	<5	<10	<10	<5	<10

TABLE XI. (Continued)

	Sample 1			Sample 2		
Element	μg/g ICP	μg/g AAS	μg/g Emission Spec Semi-Quant.	μg/g ICP	μg/g AAS	μg/g Emission Spec Semi-Quant.
Na	<50	--	<300	63	--	<300
Ni	56	61	30 - 60	26	24	15 - 30
P	<60	--	--	<60	--	--
Pb	<60	5	<30	<60	<2	<30
Pd	<8	--	--	<8	--	--
Pt	<8	--	--	<8	--	--
Rh	<8	--	--	<8	--	--
Sb	<60	--	<50	<60	--	<50
Se	<250	--	--	<250	--	--
Si	620	659	300 - 600	2541	2379	1000 - 3000
Sn	<8	--	<30	<8	--	<30
Sr	<3	--	3 - 5	<3	--	<3
Ti	56	63	30 - 60	95	104	50 - 150
V	40	39	15 - 35	19	15	10 - 20
Zn	<4	--	<50	6	--	<50
Zr	9	--	<20	6	--	<20

TABLE XII. ANALYSIS OF Al METAL STANDARDS. Background Corrector D = 2

Element	SRM L - 9 % ICP	SRM L - 9 % NBS	SRM L - 11 % ICP	SRM L - 11 % NBS	ALCOA WE-1000A % ICP	ALCOA WE-1000A % ALCOA	ALCOA SA 2650 % ICP	ALCOA SA 2650 % ALCOA
Cr	0.110	0.110	0.020	0.020	0.009	0.009	<0.0010	0.0002
Cu	0.130	0.130	0.038	0.038	0.015	0.014	0.0020	0.0021
Fe	0.490	0.500	0.570	0.570	0.014	0.014	0.0010	0.0009
Ga	--	--	--	--	0.011	0.011	0.0010	0.0013
Mg	4.140	4.150	9.210	9.200	0.002	0.002	0.0016	0.0015
Mn	0.480	0.480	0.130	0.120	0.011	0.011	0.0010	0.0009
Ni	0.086	0.085	0.020	0.020	0.0093	0.010	0.0018	0.0019
Pb	0.050	0.050	0.080	0.090	0.016	0.014	<0.0020	0.0010
Si	0.420	0.450	0.130	0.130	0.012	0.013	0.0022	0.0024
Sn	0.046	0.045	0.018	0.019	0.013	0.012	0.0020	0.0017
Ti	0.090	0.090	0.048	0.048	0.010	0.009	0.0010	0.0012
V	--	--	--	--	0.012	0.012	<0.0010	0.0003
Zn	0.510	0.510	0.026	0.025	0.015	0.014	0.0034	0.0034
Zr	0.072	0.072	--	--	--	--	--	--

ACKNOWLEDGEMENTS

I would like to thank R. Wood, R. J. Lewandowski, R. G. Parkhurst, and A. J. Ellgren for their help in this work.

INDEX